AF678067

New Advances in Oil, Gas and Geothermal Reservoirs: 2nd Edition

New Advances in Oil, Gas and Geothermal Reservoirs: 2nd Edition

Guest Editor

Daoyi Zhu

MDPI

Basel • Beijing • Wuhan • Barcelona • Belgrade • Novi Sad • Cluj • Manchester

Guest Editor
Daoyi Zhu
Faculty of Petroleum
China University of
Petroleum-Beijing at Karamay
Karamay
China

Editorial Office
MDPI AG
Grosspeteranlage 5
4052 Basel, Switzerland

This is a reprint of the Special Issue, published open access by the journal *Energies* (ISSN 1996-1073), freely accessible at: https://www.mdpi.com/journal/energies/special_issues/7G5FJI8D62.

For citation purposes, cite each article independently as indicated on the article page online and as indicated below:

Lastname, A.A.; Lastname, B.B. Article Title. *Journal Name* **Year**, *Volume Number*, Page Range.

ISBN 978-3-7258-5251-2 (Hbk)
ISBN 978-3-7258-5252-9 (PDF)
https://doi.org/10.3390/books978-3-7258-5252-9

© 2025 by the authors. Articles in this book are Open Access and distributed under the Creative Commons Attribution (CC BY) license. The book as a whole is distributed by MDPI under the terms and conditions of the Creative Commons Attribution-NonCommercial-NoDerivs (CC BY-NC-ND) license (https://creativecommons.org/licenses/by-nc-nd/4.0/).

Contents

About the Editor

Daoyi Zhu

Daoyi Zhu holds a Bachelor of Science degree in Applied Chemistry from Yangtze University (2011), followed by a master's degree in Chemical Engineering (2014) and a Ph.D. in Oil & Gas Field Development Engineering (2018) from China University of Petroleum (Beijing). His academic career includes an appointment as a visiting scholar at Missouri University of Science and Technology's Department of Petroleum Engineering (2016–2018), and he currently serves as an Associate Professor of Petroleum Engineering at China University of Petroleum-Beijing at Karamay (since 2021). As an active researcher in energy engineering, Zhu specializes in oilfield chemistry, reservoir conformance control, CCUS technologies, and geo-energy development. He has made significant scholarly contributions, including publishing 40+ peer-reviewed SCI journal papers, authoring four textbooks (*Oilfield Chemistry*, *Literature Search and Technical Writing*, *CO2 and Geo-Energy Engineering*, and *Petroleum Reservoir Engineering*), and translating multiple technical works. His editorial leadership extends to roles as Guest Editor and Executive Board Member for prominent international journals in the field.

energies

Editorial

New Advances in Oil, Gas and Geothermal Reservoirs: 2nd Edition

Daoyi Zhu

Faculty of Petroleum, China University of Petroleum-Beijing at Karamay, Karamay 834000, China; zhudaoyi@cupk.edu.cn

Oil, gas, and geothermal resources, including conventional fossil fuels (oil and natural gas) and unconventional resources (geothermal, shale gas, and tight oil), are key to meeting global energy demands [1,2]. These resources are divided into extractive (e.g., hydrocarbons) and renewable (e.g., geothermal) sources, both essential for sustaining energy systems and supporting environmental sustainability [3,4].

The exploitation of unconventional and depleted reservoirs—particularly heavy oil formations and tight reservoirs—represents a critical frontier in modern hydrocarbon production [5,6]. Improved oil recovery (IOR) techniques, such as thermal stimulation for viscous crude and advanced conformance control (e.g., water shutoff and profile modification), are essential to maximize extraction from challenging reserves with minimal energy expenditure [7–11]. Innovations like steam-assisted gravity drainage (SAGD) and cost-effective hydraulic fracturing have significantly improved the economic viability of unconventional plays [12–16]. As reservoirs age, optimizing waterflood management, reservoir characterization, and recovery rates remains a focal point of research [17–20].

The integration of big data analytics and artificial intelligence (AI) is transforming upstream operations [21,22]. Machine learning-based predictive models and real-time surveillance systems enable production optimization, hazard mitigation, and performance enhancement, while concurrently addressing challenges in reservoir stewardship and cost optimization. AI-powered decision-support tools further facilitate operational efficiency gains and lifecycle cost reductions [23,24].

Gas storage and sequestration technologies are crucial for resource management and reducing environmental impacts [25,26]. Storing natural gas and injecting CO_2 underground helps cut greenhouse gas emissions and mitigate climate change. Advanced methods for CO_2 or H_2 sequestration in depleted oil and gas fields and saline aquifers are key to meeting climate goals [27]. These technologies are evolving rapidly, with innovations in modeling, monitoring, and safety assurance.

This collection, which is in conversation with the Special Issue of *Energies*, emphasizes fundamental innovations and has compiled eight new publications on the original application of new ideas and on methodologies that will lead to new advances in oil, gas, and geothermal reservoirs.

The papers are organized into three major directions: development of mature and unconventional reservoirs, big data and artificial intelligence in oil and gas fields, and gas storage and sequestration technologies. Below, we provide an overview of the core findings from each paper, organized by these thematic categories.

The first category focuses on the challenges and innovations associated with improving recovery in both conventional and unconventional oil reservoirs, including heavy oil fields and water-flooded mature fields. Yu et al. [28] investigate the microscopic flow channels in

 https://doi.org/10.3390/en18184789

ultra-high water-cut reservoirs of the Shengli Oilfield. Their study visualizes the evolution of water flooding in reservoirs, highlighting how the dominant flow channels and residual oil distribution evolve as the water cut increases. Their research indicates that strategic adjustments in liquid extraction and flow direction can enhance oil recovery, especially in the later stages of ultra-high water-cut development, achieving an impressive recovery rate of 68.02%. This study offers valuable insights for optimizing water flooding practices in mature oilfields.

In recent years, the development of heavy oil reservoirs has gained significant attention. Tian et al. [29] investigated the changes in the thermal and physical properties of reservoir rock surfaces during the pre-heating phase of SAGD. Fu et al. [30] investigate supercritical multicomponent thermal fluids (scMCTF) for offshore heavy oil recovery. Their research reveals that the composition of thermal fluids, especially the ratio of water to organic matter, significantly impacts the recovery efficiency. By exploring reaction conditions, they find that crude oil can be effectively used instead of diesel to generate supercritical fluids, with notable economic and technical advantages. This opens new possibilities for improving thermal recovery in offshore heavy oil reservoirs. Yang et al. [31] build upon the previous study with a focus on optimizing the composition and injection rate of scMCTF for offshore heavy oil recovery. Through molecular simulation, they identify how factors such as temperature, pressure, and organic matter concentration influence the yield and composition of the thermal fluids. They propose a model for controlling the injection rate and fluid composition, providing an effective approach to enhancing the thermal recovery of heavy oil. Zhang et al. [32] examine the influence of reaction conditions on the yield of scMCTF. Their findings suggest that temperature, reaction time, and catalyst concentration have a positive correlation with product yield, while the raw material concentration negatively affects the production rate. This study contributes to the understanding of how various factors influence the efficiency of supercritical fluid generation in oil recovery.

The second category emphasizes the integration of big data and artificial intelligence (AI) technologies to optimize production in oil and gas fields, particularly through predictive modeling and fault detection. Zhang et al. [33] developed a data-driven natural gas production prediction model for volcanic reservoirs. By considering multiple factors such as formation pressure, effective reservoir thickness, and gas well production data, they establish a predictive model with high accuracy (R^2 = 0.99). This model provides valuable tools for optimizing gas production in volcanic reservoirs, enabling more efficient resource management. Zhang et al. [34] also contribute to AI applications in oil production by developing a hybrid AI model for fault prediction in rod pumping systems. Using deep learning algorithms, their model achieves a remarkable prediction accuracy of 98.61%, significantly improving the reliability and safety of rod pumping operations in Xinjiang Oilfield. The study underscores the potential of AI in enhancing production efficiency and reducing downtime due to mechanical failures.

The third category focuses on advancements in gas storage, particularly for the purposes of sequestration, which is crucial for reducing greenhouse gas emissions. Chen et al. [35] provide insights into the design and performance of a liquid helium Dewar system for thermal insulation. While not directly related to gas sequestration, the study's findings on thermal efficiency and heat leakage in cryogenic storage systems could have implications for gas storage technologies, especially in the context of developing more efficient storage solutions for gases like CO_2.

This second edition of New Advances in Oil, Gas and Geothermal Reservoirs highlights key advancements in the development of mature and unconventional (especially heavy oil) reservoirs, the application of big data and artificial intelligence in oil and gas fields, and innovative gas storage and sequestration technologies. The contributions presented in this issue offer new insights into enhancing recovery from challenging reservoirs, optimizing field operations through data-driven approaches, and advancing gas storage methods. These breakthroughs provide valuable knowledge to support the efficient and sustainable management of energy resources, guiding future research and industrial applications in the evolving energy landscape.

Conflicts of Interest: The author declares no conflicts of interest.

References

1. Guo, S.; Zhu, D. Mini-Review of Black Nanosheets for Enhanced Oil Recovery Used in Low-Permeability/Ultra-Low-Permeability Reservoirs. *Energy Fuels* **2025**, *39*, 16768–16793. [CrossRef]
2. Yang, P.; Ren, Z.-L.; Fu, J.-H.; Bao, H.-P.; Xiao, H.; Shi, Z.; Wang, K.; Zhang, Y.-Y.; Liu, W.-H.; Li, W.-H. A tectono-thermal perspective on the petroleum generation, accumulation and preservation in the southern Ordos Basin, North China. *Pet. Sci.* **2024**, *21*, 1459–1473. [CrossRef]
3. Davoodi, S.; Thanh, H.V.; Wood, D.A.; Mehrad, M.; Muravyov, S.V.; Rukavishnikov, V.S. Carbon dioxide storage and cumulative oil production predictions in unconventional reservoirs applying optimized machine-learning models. *Pet. Sci.* **2025**, *22*, 296–323. [CrossRef]
4. Zhang, H.-C.; Tang, Y.; He, Y.-W.; Qin, Y.; Luo, J.-H.; Sun, Y.; Wang, N.; Wang, D.-Q. Hydrocarbon gas huff-n-puff optimization of multiple horizontal wells with complex fracture networks in the M unconventional reservoir. *Pet. Sci.* **2024**, *21*, 1018–1031. [CrossRef]
5. Zhu, Y.-J.; Wan, Y.-Y.; Tian, Y.; Mu, H.-M.; Zhang, T.-G. Biomass thermal decomposition induced hydrogen sulfide blooming in thermal recovery reservoirs. *Pet. Sci.* **2025**, *22*, 1802–1810. [CrossRef]
6. Tan, Q.-Z.; Liu, S.-Y.; Wang, Y.-J.; Li, H.-Y.; Liu, J.-R.; Sun, W.-Y. A dual-model dual-grid upscaling method for solid-based thermal-reactive-compositional flow simulations in fractured oil shale reservoirs. *Pet. Sci.* **2025**, *22*, 2478–2492. [CrossRef]
7. Guo, S.; Cheng, H.-B.; Tan, H.-G.; Li, H.-Y.; Zhang, J.; Gao, Y.-Q.; Zhu, D.-Y. Huff-n-puff recovery performance and mechanism analysis of black nanosheets in low-permeability reservoirs based on NMR technology. *Pet. Sci.* **2025**, *22*, 2992–3004. [CrossRef]
8. Zhu, D.; Guo, S.; Lu, J.; Yang, Y.; Zhang, T.; Zhang, J.; Gao, Y.; Tan, H.; Cheng, H.; Li, H. Influence of Injection Parameters of Black Nanosheets on Enhancing Oil Recovery Performance in Low-Permeability Reservoirs. *Energy Fuels* **2025**, *39*, 6241–6250. [CrossRef]
9. Ali, A.B.; Hamza, A.; Almakimi, A.A.; Saad, M.; Hussein, I.A.; Bai, B. Carboxymethyl cellulose-based preformed particle gels for water management in oil and gas reservoirs. *Geoenergy Sci. Eng.* **2024**, *241*, 213164. [CrossRef]
10. Elaf, R.; Hamza, A.; Nimir, H.; Saad, M.; Hussein, I.A.; Bai, B. Development of inorganically cross-linked Sulfonated Polyacrylamide Preformed particle gels for Conformance Control: Impact of anionic groups. *Energy Fuels* **2024**, *38*, 2883–2897. [CrossRef]
11. Alotibi, A.; Song, T.; Bai, B.; Schuman, T. Transport and Plugging Performance Evaluation of a Novel Recrosslinkable Microgel Used for Conformance Control in Mature Oil Fields with Superpermeable Channels. *SPE J.* **2025**, *30*, 823–835. [CrossRef]
12. Wu, J.-X.; Li, S.-F.; Li, Q.-F.; Yan, F.; Zhou, Q.-L.; Ma, S.; Zhang, Y.-H.; Zhao, S.-Q.; Shi, Q. Characterization of chemical composition of high viscosity heavy oils: Macroscopic properties, and semi-quantitative analysis of molecular composition using high-resolution mass spectrometry. *Pet. Sci.* **2024**, *21*, 3612–3620. [CrossRef]
13. Lei, X.-T.; Ahmadi, M.; Chen, Z. Molecular insights into oil detachment from hydrophobic quartz surfaces in clay-hosted nanopores during steam–surfactant co-injection. *Pet. Sci.* **2024**, *21*, 2457–2468.
14. BinDahbag, M.; Atta, D.; Bagherzadeh, H.; Nath, D.; Ateeq, M.; Kheirollahi, S.; Bamzad, S.; Turkman, S.; Kamal, B.; Hassanzadeh, H. Effectiveness of Natural Gas Condensate as a Viable Solvent in ES-SAGD Processes: An Experimental Investigation Using a 3-D Physical Model. *Energy Fuels* **2024**, *38*, 16036–16048. [CrossRef]
15. Yang, S.; Huang, S.; Jiang, Q.; Jiang, G.; Zhou, X.; Yu, C. An innovative in situ solvent generation enhanced SAGD technique: Mechanism analysis based on numerical simulation. *Fuel* **2024**, *364*, 131020. [CrossRef]
16. Atta, D.; BinDahbag, M.; Bagherzadeh, H.; Nath, D.; Shah Bukhari, S.S.U.; Turkman, S.; El-Shazly, A.; Hassanzadeh, H. A 3-D Physical Model Experimental Study of the ES-SAGD Process Utilizing Dimethyl Ether (DME) as a Solvent. *Energy Fuels* **2025**, *39*, 15672–15688. [CrossRef]
17. Halari, D.; Yadav, S.; Kesarwani, H.; Saxena, A.; Sharma, S. Nanoparticle and surfactant stabilized carbonated water induced in-situ CO2 foam: An improved oil recovery approach. *Energy Fuels* **2024**, *38*, 3622–3634. [CrossRef]
18. Geng, W.; Dong, B.; Zhang, Y.; Li, C.z.; Yang, N.; Si, Y.; Dai, C.; Zhao, G.; Cheng, H. A Temperature-Salinity Responsive MCPs System for Water Shutoff in Tight Gas Reservoirs. *Energy Fuels* **2025**, *39*, 16830–16843. [CrossRef]

19. Song, T.; Bai, B.; Huang, R.; Zhang, S.; Liu, P.; Eriyagama, Y.; Tian, X.; Ahdaya, M.; Schuman, T. Development and evaluation of lysine-crosslinked re-crosslinkable particle gel for water control in high-temperature reservoirs. *J. Mol. Liq.* **2024**, *407*, 125133. [CrossRef]
20. Ben Ali, A.; Elaf, R.; Almakimi, A.A.; Saad, M.; Hussein, I.A.; Bai, B. Development of agar-based preformed particle gel for water control in high-salinity reservoirs. *Ind. Eng. Chem. Res.* **2024**, *63*, 8524–8541. [CrossRef]
21. Peng, C.; Zhang, H.-L.; Fu, J.-H.; Su, Y.; Li, Q.-F.; Yue, T.-Q. A novel drilling parameter optimization method based on big data of drilling. *Pet. Sci.* **2025**, *22*, 1596–1610. [CrossRef]
22. Kang, M.-L.; Zhou, J.; Zhang, J.; Xiao, L.-Z.; Liao, G.-Z.; Shao, R.-B.; Luo, G. An integrated method of data-driven and mechanism models for formation evaluation with logs. *Pet. Sci.* **2025**, *22*, 1110–1124. [CrossRef]
23. Mkono, C.N.; Chuanbo, S.; Mulashani, A.K.; Abelly, E.N.; Kasala, E.E.; Shanghvi, E.R.; Emmanuely, B.L.; Mokobodi, T. Improved Reservoir Porosity Estimation Using an Enhanced Group Method of Data Handling with Differential Evolution Model and Explainable Artificial Intelligence. *SPE J.* **2025**, *30*, 1922–1940. [CrossRef]
24. Meng, H.; Lin, B.; Jin, Y. Stop Using Black-Box Models: Application of Explainable Artificial Intelligence for Rate of Penetration Prediction. *SPE J.* **2024**, *29*, 6640–6654. [CrossRef]
25. Zhu, D.; Zhao, Q.; Chen, P.; Lu, J.; Yang, Y.; Guo, S.; Zhang, T. Laboratory Evaluation of Antileakage Performance against CO_2 of Alkali-Activated Gel-Reinforced Cement for Carbon Capture, Utilization, and Storage. *SPE J.* **2025**, *30*, 3776–3791. [CrossRef]
26. Su, D.; Mao, T.; Huang, S.; Li, Z.; Zhou, C. Novel Prediction Method for the Service Life of the Cement Sheath in Gas Storage Wells. *SPE J.* **2025**, *30*, 3437–3455. [CrossRef]
27. Wang, Y.; Jin, Y.; Pang, H.; Lin, B. Upscaling for Full-Physics Models of CO_2 Injection Into Saline Aquifers. *SPE J.* **2025**, *30*, 3065–3082. [CrossRef]
28. Yu, C.; Zhang, M.; Chen, W.; Zhang, S.; Wang, S. A Microscopic Experimental Study on the Dominant Flow Channels of Water Flooding in Ultra-High Water Cut Reservoirs. *Energies* **2024**, *17*, 5756. [CrossRef]
29. Tian, J.; Huang, S.; Dong, M.; Yan, W.; Qi, Z. A Study on the Thermal Physical Property Changes in Formation Rocks during Rapid Preheating of SAGD. *Energies* **2024**, *17*, 3834. [CrossRef]
30. Fu, Q.; Tian, J.; Liu, Y.; Qi, Z.; Jiao, H.; Yang, S. Comparison of the Reaction Characteristics of Different Fuels in the Supercritical Multicomponent Thermal Fluid Generation Process. *Energies* **2024**, *17*, 5376. [CrossRef]
31. Yang, S.; Qi, Z.; Tian, J.; Dong, M.; Zhang, W.; Yan, W. Composition and Injection Rate Co-Optimization Method of Supercritical Multicomponent Thermal Fluid Used for Offshore Heavy Oil Thermal Recovery. *Energies* **2024**, *17*, 5239. [CrossRef]
32. Zhang, W.; Qi, Z.; Tian, J.; Xu, F.; Kong, D.; Dong, M.; Yang, S.; Yan, W. Influence of Reaction Conditions on the Yield of Supercritical Multicomponent Thermal Fluids. *Energies* **2024**, *17*, 5012. [CrossRef]
33. Zhang, H.; Pu, J.; Zhang, L.; Deng, H.; Yu, J.; Xie, Y.; Tong, X.; Man, X.; Liu, Z. Gas Production Prediction Model of Volcanic Reservoir Based on Data-Driven Method. *Energies* **2024**, *17*, 5461. [CrossRef]
34. Zhang, A.; Zhao, Y.; Li, X.; Fan, X.; Ren, X.; Li, Q.; Yue, L. Development of a Hybrid AI Model for Fault Prediction in Rod Pumping System for Petroleum Well Production. *Energies* **2024**, *17*, 5422. [CrossRef]
35. Chen, Y.; Guo, L.; Jia, Q.; Xie, X.; Zhu, W.; Wang, P. Research on the Liquid Helium Insulation Characteristics of an Experimental System. *Energies* **2025**, *18*, 1349. [CrossRef]

Disclaimer/Publisher's Note: The statements, opinions and data contained in all publications are solely those of the individual author(s) and contributor(s) and not of MDPI and/or the editor(s). MDPI and/or the editor(s) disclaim responsibility for any injury to people or property resulting from any ideas, methods, instructions or products referred to in the content.

energies

MDPI

Article

A Microscopic Experimental Study on the Dominant Flow Channels of Water Flooding in Ultra-High Water Cut Reservoirs

Chunlei Yu [1,*], Min Zhang [1], Wenbin Chen [2], Shiming Zhang [1] and Shuoliang Wang [2]

1 Research Institute of Exploration and Development, Shengli Oilfield Company, Sinopec Corporation, Dongying 257001, China; zhangmin658.slyt@sinopec.com (M.Z.); zhangshm855.slyt@sinopec.com (S.Z.)
2 College of Energy Resources, China University of Geosciences (Beijing), Beijing 100190, China; cwbscience@163.com (W.C.); wangshuoliang@cugb.edu.cn (S.W.)
* Correspondence: yuchunlei.slyt@sinopec.com

Abstract: The water drive reservoir in Shengli Oilfield has entered a stage of ultra-high water cut development, forming an advantageous flow channel for the water drive, resulting in the inefficient and ineffective circulation of injected water. Therefore, the distribution characteristics of water drive flow channels and their controlled residual oil in ultra-high water cut reservoirs are of great significance for treating water drive dominant flow channels and utilizing discontinuous residual oil. Through microscopic physical simulation of water flooding, color mixing recognition and image analysis technology were used to visualize the evolution characteristics of water flooding seepage channels and their changes during the control process. Research has shown that during the ultra-high water content period, the shrinkage of the water drive seepage channel forms a dominant seepage channel, forming a "seepage barrier" at the boundary of the dominant seepage channel, and dividing the affected area into the water drive dominant seepage zone and the seepage stagnation zone. The advantage of water flooding is that the oil displacement efficiency in the permeable zone is as high as 80.5%, and the remaining oil is highly dispersed. The water phase is almost a single-phase flow, revealing the reason for high water consumption in this stage. The remaining oil outside the affected area and within the stagnant flow zone accounts for 89.8% of the remaining oil, which has the potential to further improve oil recovery in the later stage of ultra-high water cut. For the first time, the redundancy index was proposed to quantitatively evaluate the control effect of liquid extraction and liquid flow direction on the dominant flow channels in water flooding. Experimental data showed that both liquid extraction and liquid flow direction can regulate the dominant flow channels in water flooding and improve oil recovery under certain conditions. Microscopic physical simulation experiments were conducted through the transformation of well network form in the later stage of ultra-high water content, which showed that the synergistic effect of liquid extraction and liquid flow direction can significantly improve the oil recovery effect, with an oil recovery rate of 68.02%, deepening the understanding of improving oil recovery rate in the later stage of ultra-high water content.

Keywords: ultra-high water cut period; advantageous seepage channels; seepage barrier; remaining oil distribution; improved oil recovery

1. Introduction

The onshore oil reservoirs in eastern China are mainly composed of clastic rock deposits, and the reservoirs have complex heterogeneity in both horizontal and vertical directions. The reservoir development method is mainly the water drive. At present, it has generally entered a high water content stage, with a water content of more than 90%, but the average recovery rate is less than 40%, and more than 60% of the oil and gas resources are retained underground. In the high water content stage, various dominant seepage channels are generally formed in the main reservoirs [1,2], and the injected water circulates ineffectively in the reservoir, which inhibits the potential of the remaining oil in other layers [3–5], resulting in the deterioration of the output and economic benefits of oilfields.

In oilfields, through conventional means such as tracer monitoring and isotope analysis, it can be determined whether there is a certain dominant flow channel between the injection well and the production well. However, when multiple reservoirs are developed at the same time, it is difficult to determine which layer is the dominant flow channel [6]. In addition, the dominant seepage channel technology based on reservoir numerical simulation, streamline simulation, and reservoir engineering methods has also been widely developed. The accuracy of reservoir numerical simulation in identifying the dominant seepage path depends largely on rock physical parameters such as grid porosity, permeability, and relative permeability. The simulation process obtains these parameters through wells' point data difference, which does not allow for an accurate understanding of the reservoir [7]. Streamline simulation technology can effectively model and predict the dominant flow path, but due to the uncertainty of mapping physical properties from grid to one-dimensional space, certain errors may occur in the calculation [8]. Based on reservoir engineering methods, the quantitative evaluation of inter-well dominant flow paths has attracted widespread attention due to the advantages of easy data acquisition, rich information, and low cost. Commonly used methods include excess water analysis [9], multivariate linear regression model [10], capacitance-resistance model (CRM) [11,12], and inter-well connectivity numerical simulation model (INSIM) [13–15]. The excess water analysis method [9] is based on traditional seepage theory, based on which excess water (the difference between the actual injected water volume and the predicted injected water volume) is divided into different injection and production directions and reservoir parameters such as dominant flow, permeability, and pore throat radius are determined. It assumes that fluid flow follows Darcy's law and regards the flow in the dominant channel as a one-dimensional flow. However, a large number of experiments have shown that high-speed non-Darcy flow is widely observed in the dominant channel, indicating that there is a significant difference between assumptions and actual situations, so the results obtained by excess water analysis are less reliable. The multivariate linear regression method [10] estimates the weight coefficient of the degree of the well connectivity through the volatility of well production data but ignores the attenuation and hysteresis behavior of injected water between well pairs, which usually leads to some defects in practical applications. The capacitance-resistance model [12,16] takes into account the attenuation and hysteresis of the inter-well signal and derives the relationship between the production well fluid production, initial production, injection water volume, and bottom hole pressure based on the material balance equation. There are still some assumptions in CRM, and whether these assumptions are valid in oilfields has not been confirmed. The INSIM is based on the principle of material balance, based on which a numerical simulation kernel function of the well connectivity (such as a capacitance-resistance model) is constructed [17,18], simplifying the kernel function with the help of predetermined assumptions, and then the dominant flow problem is solved. The above methods all use field tests and simulation calculation methods to study the dominant channels of water flow.

In order to explore the evolution mechanism of the dominant flow channel and its influence on reservoir parameters, a large number of indoor experimental studies have been conducted. Kalaydjian [19] studied the microscopic residual oil distribution when oil, gas, and water coexisted by scanning electron microscopy; in a series of long-term water injection experiments, Liu [20] found that with an increase in water injection pore volume, the reservoir permeability polarized, the permeability of the high-permeability area increased, and the permeability of the low-permeability area decreased. The clay mineral content has the greatest influence on changes in rock physical parameters, while the pore throat distribution and rock wettability have little influence. Liu [21] revealed that changes in porosity, permeability, heterogeneity, and cementation are the main geological factors explaining the formation of the dominant flow channel. Zhong [22] found that changes in porosity, permeability, heterogeneity, and cementation are the main geological factors explaining the formation of the dominant flow channel. It is believed that the sedimentary microfacies and sand body thickness at the macroscale control the formation of the dominant flow channel, while the sand body lithology, stratigraphic rhythm, rock physical properties, and diagenetic evolution degree at the microscale usually have a greater impact on the formation of the dominant flow channel. Yao [23] found that injected water can flush the surface of oil-bearing pores, causing clay minerals to swell and migrate and thus resulting in the thinning or destruction of the oil film attached to the clay surface, thereby changing the wettability of the reservoir rock. Huang [24] believed that in medium- and low-permeability reservoirs, after long-term water injection development, the porosity and permeability increase, the heterogeneity intensifies, the water sensitivity decreases, and the acid sensitivity intensifies.

Indoor experiments have shown that by improving the hydrodynamic conditions, the water drive efficiency can exceed 70%, and the recovery rate can reach more than 50% [9,25,26]. However, the water flooding effect observed in large-scale pilot experiments and indoor simulations is still challenging [27–30]. Among these challenges, understanding the evolution of water flooding channels during the regulation process and their effective characteristics, as well as the distribution and utilization characteristics of remaining oil under the control of water flooding channels, are key issues in improving the effectiveness of hydrodynamic regulation in the ultra-high water cut period [6,31,32]. Previous studies lacked research on the dominant flow channel system from the perspective of fluid mechanics. There is an urgent need to establish a new method for identifying and observing the dominant flow channels in ultra-high water cut reservoirs that takes into account the time-varying effects of reservoir parameters and uses it to effectively guide the actual development of oilfields. To this end, we studied the evolution characteristics of seepage channels during water flooding through microscopic physical simulation experiments and visualized the dominant seepage channels of water flooding in the late stage of ultra-high water cut; the color mixing principle was used to change the color of the injected water at different stages of water flooding to trace the water seepage channels. For the first time, it was discovered that the water flooding seepage channels in the late stage of ultra-high water cut shrank and solidified to form dominant seepage channels, and the control effect of the dominant seepage channels on the distribution of residual oil was quantitatively analyzed. Furthermore, a redundant index was established to quantitatively evaluate the inhibitory effect of two hydrodynamic control measures, namely, liquid extraction and liquid flow diversion, on the dominant seepage channels of water flooding. Finally, microscopic physical simulation of chemical flooding and hydrodynamic control was used to synergistically break the existing dominant seepage channels of water flooding, reconstruct an efficient displacement flow field, greatly improve the degree of remaining oil utilization, and achieve a recovery rate of more than 70%, thus achieving the goal of significantly improving the recovery rate in the late stage of ultra-high water cut.

2. Microscopic Simulation Methods for Waterflooding Microflow in Porous Media

Microscopic simulation of waterflooding microflow is one of the important means to study the characteristics of waterflooding flow in porous media. The entire physical simulation method includes three key technologies: microscopic model fabrication, experimental control techniques for microscopic waterflooding, and post-processing techniques for microscopic waterflooding image data. Among them, the fabrication of the microscopic model is key.

During the experimental process, thin sections of representative reservoir cores were selected, and binary images of the pore network were extracted. Using image fusion and stitching methods, multiple binary images of pore networks were reconstructed into pore network patterns of different sizes and specifications. Then, the pore network patterns were mapped and etched onto a glass substrate using laser etching technology. A precision engraving machine was used to drill holes based on the well-pattern deployment, simulating well positions. Finally, the etched glass substrate was encapsulated using high-temperature bonding technology.

The process of microscopic waterflooding experiments was the same as that of conventional core waterflooding experiments. A high-resolution camera captured and stored dynamic and static images of the waterflooding process in a computer. The experimental procedure is shown in Figure 1. After the experiment, image analysis techniques were used to analyze and quantitatively calculate the image data qualitatively [33]. This analysis clarifies the waterflooding process's microscopic flow characteristics, the oil displacement's effectiveness, and the state of remaining oil distribution. It provides theoretical support for regulating water flow paths during the ultra-high water cut period, the effective mobilization of remaining oil, and improving oil recovery.

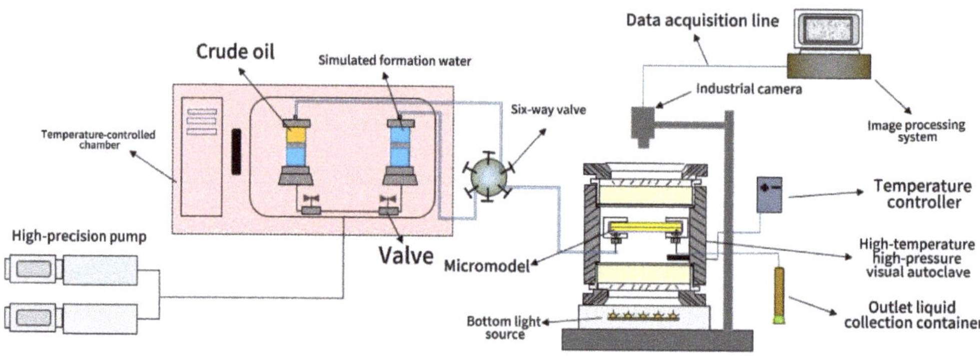

Figure 1. Microscopic oil displacement simulation's experimental device.

3. Characteristics of the Evolution Process of Dominant Seepage Channels in Water Flooding

3.1. Characteristics of the Evolution Process of Seepage Channels Based on Injected Water Color Tracing

In reservoir water flooding development, the oil phase is continuously cut by the water phase, increasing dispersion. The two-phase flow of oil and water gradually changes from continuous flow to discontinuous flow. During the water flooding process, due to the increasing dynamic heterogeneity of reservoir properties and residual oil saturation, the distribution of residual oil becomes more scattered and smaller in scale. The distribution of the seepage channels for the injected water tends to stabilize, eventually entering the ultra-high water cut stage. Laboratory microphysical simulation visual studies were used to study the water flooding oil process. Initially, blue water was used for water flooding until the ultra-high water cut stage, and then pink water was used to displace the crude oil, as shown in Figure 2.

From Figure 2, it can be observed that in the low water cut stage, due to reservoir heterogeneity, water injection leads to fingering (Figure 2a). In the medium water cut stage, due to the interference between the two-phase flow of oil and water in the porous medium, the swept area continues to expand even after the injected water breaks through to the outlet (Figure 2b). In the high water cut stage, the phase interference in porous media is weakened, and the spreading ability of injected water is weakened with the increase in outlet moisture content (Figure 2c). In the ultra-high water cut stage, the swept area reaches its maximum, and residual oil distribution becomes relatively stable (Figure 2d).

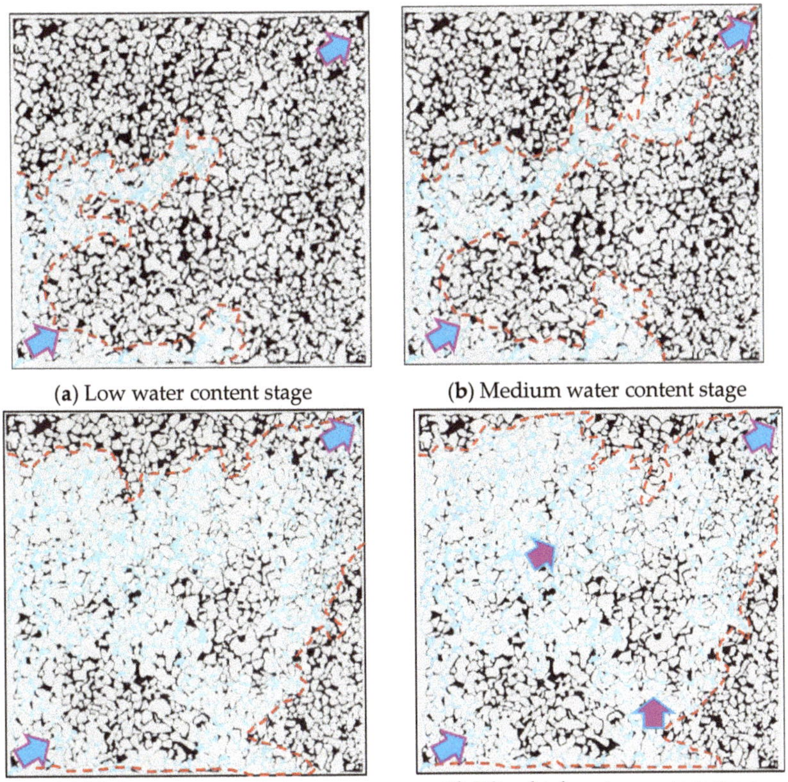

(a) Low water content stage (b) Medium water content stage (c) High water content stage (d) Ultra-high water content stage

Figure 2. Evolution process of seepage path from water drive to high water content stage. The arrows represent the direction of flow.

3.2. Visualization and Identification of Dominant Seepage Channels in Water Flooding Based on Color Mixing Principles

After entering the ultra-high water cut stage, the injected water changed from blue to pink. During the experiment, as the pink water gradually entered the porous medium, it mixed with the blue water in the pores. This mixing process is shown in Figure 3a–d, with local areas of the images enlarged. Figure 3d shows a microscopic displacement image after injecting 30 pore volumes (PV) of pink water, representing the late stage of ultra-high water cut, with full-color mixing.

Figure 3 shows that the injected water changed from blue to pink during the ultra-high water cut stage. During the injection process, a color mixing process occurred between the blue and pink water, with the colors transitioning from blue to light purple to purple to pink. Different colors represent different intensities of color mixing, indicating the extent of mixing caused by the seepage intensity of the injected water. In the late stage of the ultra-high water cut stage, the color mixing process reached equilibrium, and the pink areas in Figure 3d show the complete replacement of blue water with pink water. This seepage area represents the dominant seepage channels during the late stage of ultra-high water cuts.

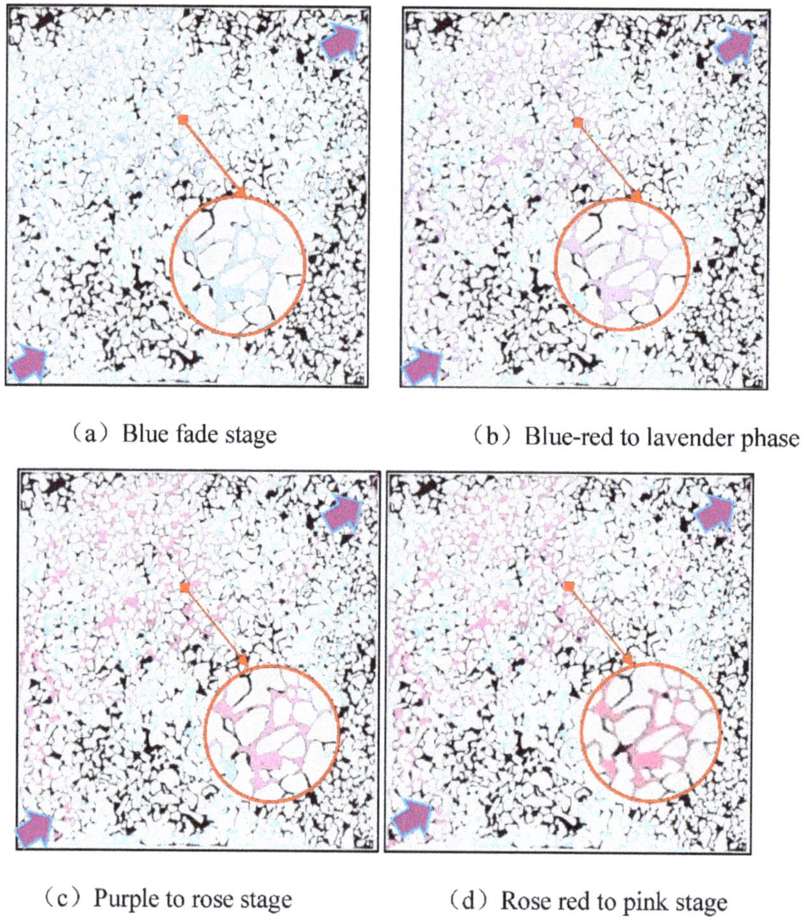

Figure 3. Tracing process of dominant water drive seepage channels in the later stage of ultra-high water content. The arrows represent the direction of flow.

In summary, the evolution characteristics of dominant seepage channels during water flooding include directional advancement in the low water cut stage, lateral expansion in the medium–high water cut stage, and contraction and stabilization in the ultra-high water cut stage. The tracing process of dominant seepage channels during water flooding indicates that in the late stage of ultra-high water cut, flow mainly occurs in some pores within the swept area, rather than all pores participating in the seepage flow.

4. The Control Effect of Water Flooding Dominant Flow Channels on Remaining Oil

4.1. Distribution Characteristics of Flow Regions Controlled by Dominant Flow Channels in Water Flooding

The water flooding process involves a two-phase flow, where the porous medium contains either water or oil. The distribution and displacement intensity of water flooding flow channels control the remaining oil. Figure 4 shows the distribution of oil and water under the control of dominant flow channels in the late stage of high water cut.

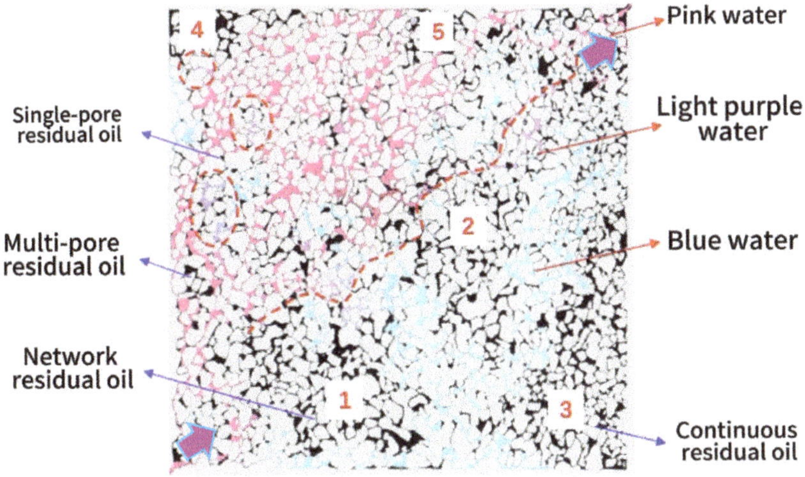

Figure 4. Distribution of remaining oil under the control of water flow advantage channels. Numbers 1–5 label the five areas filled with remaining oil. The blue arrows label each of the four remaining oil types. Red arrows mark the color of the liquid.

As shown in Figure 4, the sweep efficiency of water flooding in the late stage of high water cut is relatively high, with the swept area accounting for 85.4%. The swept area is further divided into the dominant flow area (pink, 42.7%) and the stagnant flow area (blue, 42.7%). In the late stage of high water cut, water in the stagnant flow area remains blue, indicating that the pores in this area do not participate in flow, no mixing process occurs, and blue water is not replaced with pink water. The pores in the dominant flow area are the main flow channels, and the water in these pores is completely replaced with pink water. Further observation of Figure 4 reveals a relatively continuous light purple mixed-color strip (near the red ant lines) at the boundary between the dominant flow area and the stagnant flow area. This strip acts as a flow barrier, separating the dominant and stagnant flow areas.

4.2. Oil Displacement Efficiency and Remaining Oil Analysis in Different Flow Areas

The regional distribution of water flooding flow paths in the late stage of high water cut controls the remaining oil distribution. The type and scale of remaining oil vary significantly across different flow areas. In the unswept area, the remaining oil is distributed in contiguous patches (marked as locations 3, 4, and 5 in Figure 4). In the stagnant flow area, the remaining oil is mainly bypassed network-like oil (marked as locations 1 and 2 in Figure 4), with some single-pore and multi-pore dispersed remaining oil. In the dominant flow area, the remaining oil is highly dispersed, mainly consisting of single-pore and multi-pore dispersed oil. The percentage of the remaining oil in different flow areas relative to the total remaining oil and the oil displacement efficiency is shown in Figure 5.

As shown in Figure 5, the oil displacement efficiency in different flow areas during the late stage of high water cut follows the following order: dominant flow area (80.5%) > stagnant flow area (44.6%) > unswept area (0). The proportion of remaining oil in each flow area relative to the total remaining oil is as follows: stagnant flow area (49.5%) > unswept area (40.3%) > dominant flow area (10.2%). The dominant flow area, serving as the main flow channel, contains only 10% of the remaining oil but has an oil displacement efficiency of up to 80.5%. In the experiments, the water phase in this area approached single-phase flow during the late stage of high water cut, resulting in the lowest flow resistance. The stagnant flow area, with nearly 50% of the remaining oil, is crucial for tapping the remaining oil in the late high water cut stage.

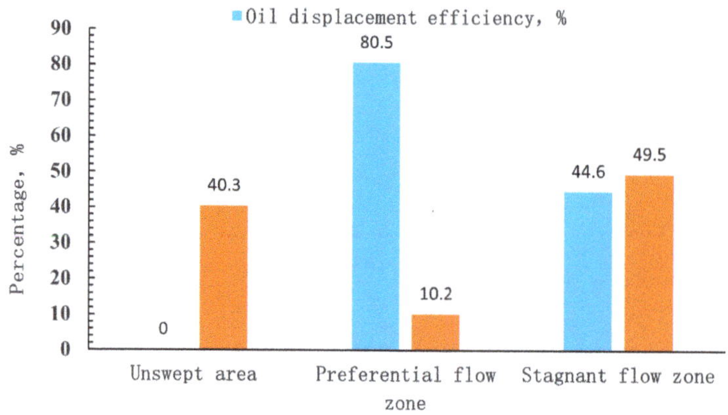

Figure 5. Oil displacement efficiency and remaining oil proportion in different seepage areas of water flooding.

In summary, the water flooding flow areas in the late stage of high water cut are divided into unswept areas, dominant flow areas, and stagnant flow areas. The remaining oil in the dominant flow area is highly dispersed and small in scale, with the water phase approaching single-phase flow, making the recovery of the remaining oil challenging. The stagnant flow area contains bypassed network-like remaining oil, accounting for nearly 50% of the total remaining oil, making it a key area for tapping remaining oil in the late stage of high water cut. The unswept area contains 40.3% of the total remaining oil and is the most difficult to mobilize under unchanged hydrodynamic conditions.

5. Regulation and Evaluation of Water Flooding Dominant Permeability Channels

In the late stage of the ultra-high water cut, dominant permeability channels are developed in the swept areas, which control the distribution of remaining oil through the barrier effect of these channels. The field regulates these dominant water flooding channels by increasing liquid production or altering liquid flow direction, further enhancing recovery efficiency. Microscopic physical simulations in the laboratory were used to study the effects of increasing liquid production and altering liquid flow direction on the regulation of dominant water flow channels. In the experiment, colorless water was first used to drive to the ultra-high water cut stage, and then blue water was used to increase liquid production for displacement gradually, and finally, pink water was used for flow direction alteration displacement. Image analysis methods were employed to quantitatively evaluate the effects of increasing liquid production and altering liquid flow direction on the regulation of dominant water flow channels.

5.1. Characteristics and Effect Evaluation of Liquid Production Regulation of Water Flooding Permeability Channels

Increasing liquid production by increasing the displacement pressure leads to the non-selective expansion of the swept area of the injected water, thereby inhibiting the dominant permeability channels in the late stage of ultra-high water cut and further enhancing recovery efficiency. The image of the experiment for tracing the dominant permeability channels at 1 μL/min water flooding until reaching the ultra-high water cut stage is shown in Figure 6.

Figure 6 shows that in the late stage of ultra-high water cut, the injected water generally sweeps through the area, developing dominant water flooding permeability channels (blue) within the swept region. The experimental images were re-rendered to quantitatively evaluate the effect of liquid production expansion using image analysis methods, as shown in Figure 7a. To illustrate the evolution of dominant water flooding permeability channels during the stepwise liquid production process, the re-rendered images of these channels are shown in Figure 7b–d.

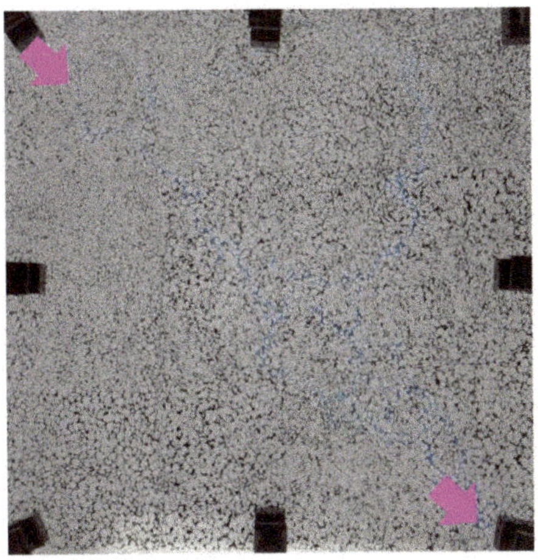

Figure 6. Tracer image of the dominant seepage channel in the water drive during the later stage of ultra-high water content. The arrows represent the direction of flow.

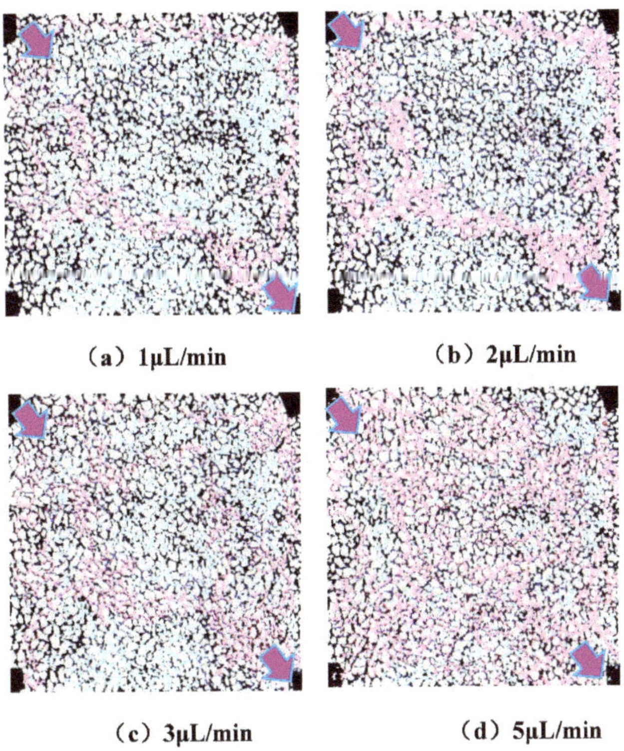

Figure 7. Evolution of dominant seepage channels during stepwise liquid extraction. The arrows represent the direction of flow.

Figure 7 shows that with the increase in injection speed, the dominant water flooding permeability zone (pink) gradually overcomes the barrier effect and continues to expand. In contrast, the stagnant flow zone continuously shrinks. On the pore scale, the number of pores participating in the flow within the swept area gradually increases. The remaining pores that are not yet involved in the flow are defined as redundant pores. The redundancy index is defined to quantitatively evaluate the effect of liquid production regulation on the dominant water flooding permeability zone:

$$RI = \left(1 - \frac{\text{Dominant seepage area}}{\text{Affected area}}\right) \times 100\%$$

The relationship between the increased injection speed and the redundancy index in the late stage of ultra-high water cut is shown in Figure 7.

Figures 7 and 8 show that in the late stage of ultra-high water cut, with the stepwise increase in liquid production, the redundancy index continuously decreases, the dominant permeability zone continuously expands, the barrier effect is continuously broken, and oil recovery efficiency increases. Within the experimental range, when the injection speed is increased fivefold (to 5 μL/min), the redundancy index is 22.8%. The primary effect of increased liquid production is to mobilize the remaining oil in the redundant zone, achieving an oil recovery efficiency of 46.7%. This is mainly due to the non-selective flow of injected water breaking the original barrier effect, expanding the dominant permeability zone, and displacing the remaining oil in redundant pores, thus improving oil recovery efficiency.

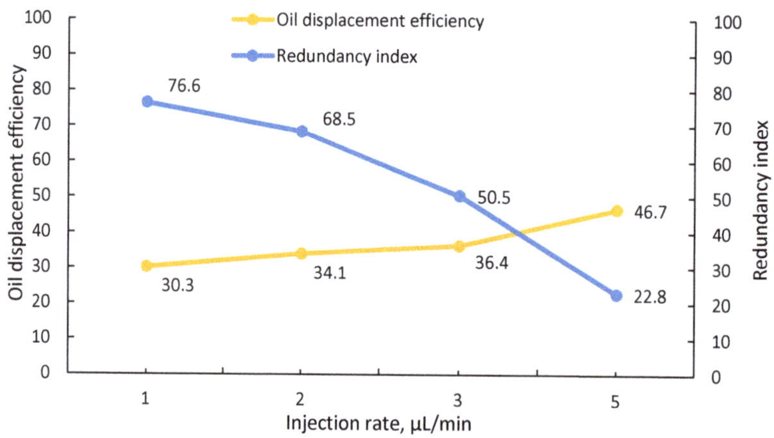

Figure 8. Relationship between stepwise liquid extraction and redundancy index in the later stage of ultra-high water content.

5.2. Characteristics and Effect Evaluation of Flow Direction Alteration in Regulating Water Flooding Permeability Channels

After the blue water flooding and stepwise liquid production equilibrium, pink water was used to alter the flow direction by 90° for displacement, regulating the water flooding permeability channels. The image of the experimental process is shown in Figure 9a. After the equilibrium of flow direction alteration displacement, blue water was used for flow path tracing, and its image is shown in Figure 9b.

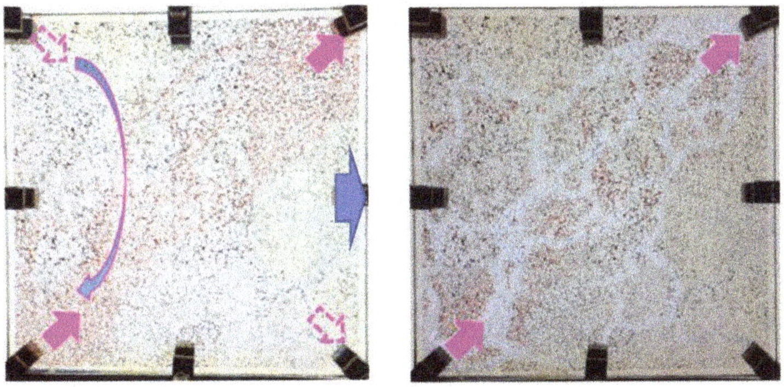

(a) (b)

Figure 9. Tracer images of fluid flow turning in the water drive advantage seepage channel. (**a**) The image of the experimental process of liquid flow turns. (**b**) Tracer image of fluid flow reaching equilibrium in the dominant seepage channel. Solid pink arrows represent the main flow direction. Dashed arrows represent flow directions perpendicular to the main flow direction. Short blue arrows represent the direction of flow after outflow. The long blue arrow represents the direction of flow after improvement.

Figure 9 shows that the backward water flooding sewerage channel is reconstructed by liquid flow, and Figure 10 shows the corresponding image after secondary rendering.

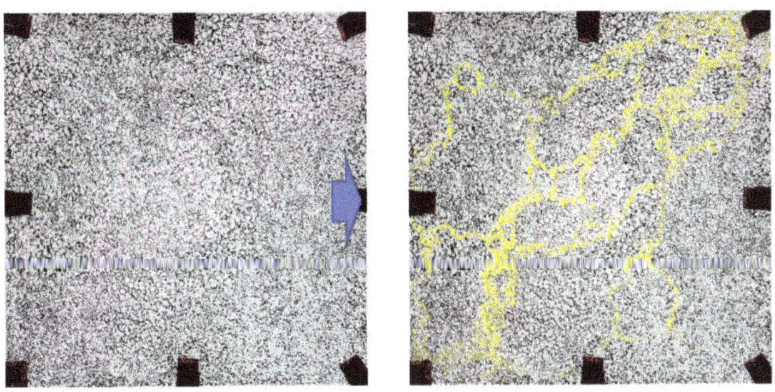

Figure 10. Evolution of dominant seepage channels during fluid flow turning. Short blue arrows represent the direction of flow.

It can be seen from Figure 10 that after the liquid flow reaches displacement equilibrium, there are still dominant seepage channels of water flooding in the reconstructed water flooding seepage channels (yellow area in Figure 10), as indicated by quantitative analysis of redundancy index and displacement efficiency (see Figure 11).

From Figure 11, it can be seen that in the initial stage of flow direction alteration, oil recovery efficiency significantly increases. As displacement continues, dominant water flooding permeability channels form again, the redundancy index increases, and oil recovery efficiency almost no longer increases, entering a high water consumption displacement stage. The changes in the dominant water flooding permeability channels and oil recovery efficiency after flow direction alteration in the late stage of ultra-high water cut indicate that the effectiveness of flow direction alteration in enhancing recovery is time-sensitive. In the ultra-high water cut stage, dominant water flow channels re-form, reducing the effectiveness of flow direction alteration in improving oil recovery.

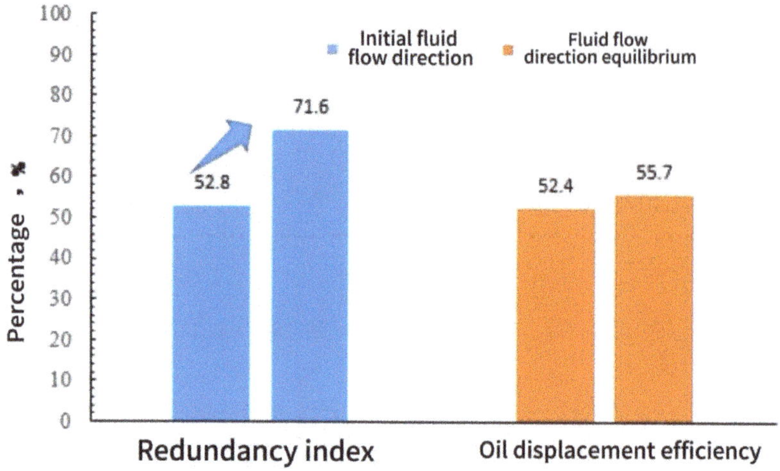

Figure 11. Changes in the redundancy index and oil displacement efficiency of fluid flow direction displacement until reaching equilibrium.

5.3. Evaluation of the Effect of Synergistic Regulation of Water Flooding Permeability Channels on Enhanced Oil Recovery

In the late stage of ultra-high water cut, after multiple rounds of increased liquid production and flow direction alteration, the remaining oil becomes highly dispersed, and the adaptability of single water flooding regulation technology gradually deteriorates. Indoor experiments were conducted to simulate the synergistic effect of increased liquid production and flow direction alteration on enhancing recovery through flow field regulation. As shown in Figure 12, a one-injection and one-production microphysical simulation experiment was first conducted. After displacement to the ultra-high water cut stage, the well pattern was changed to a two-injection and six-production pattern while maintaining the injection speed, achieving the synergistic regulation of the flow field through increased liquid production and flow direction alteration. The results of the sweep efficiency and oil recovery efficiency analyses after equilibrium in the one-injection and one-production experiments, as well as the two-injection and six-production experiments, are shown in Figure 13.

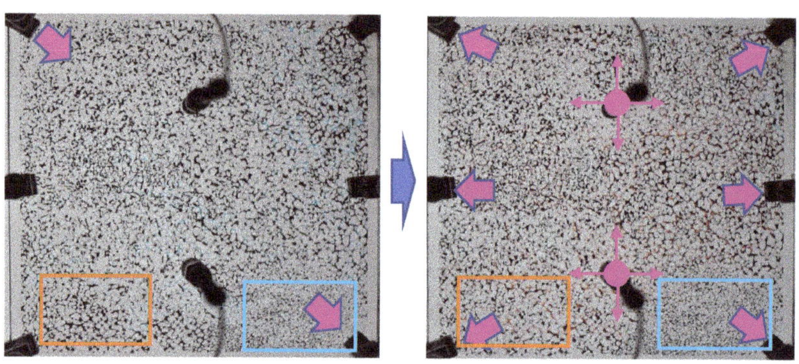

Figure 12. Collaborative regulation of water drive's seepage channels through well network transformation. The pink arrow represents the direction of fluid flow. The pink origin represents the location of the simulated injection well. The orange and blue boxes point out two areas where the fluid flow conditions differ.

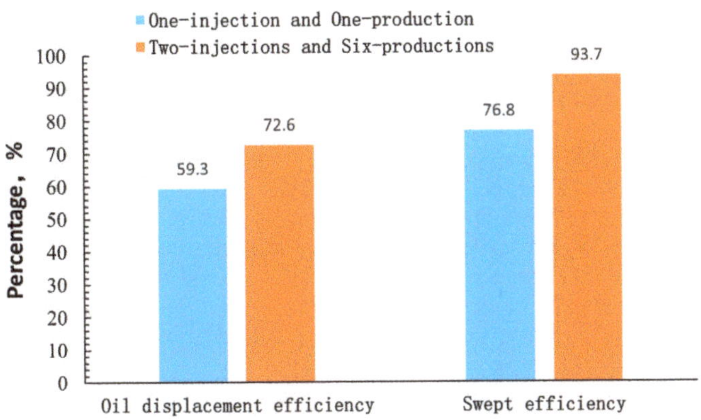

Figure 13. Effect of well network transformation on oil displacement efficiency and sweep coefficient.

Figure 13 shows that after changing from a one-injection and one-production to a two-injection and six-production method, the sweep efficiency increased from 76.8% to 93.7%, and oil recovery efficiency increased from 59.3% to 72.6%. Two typical regions were selected for comparative analysis in the model to further analyze the improvement in oil recovery efficiency by increased liquid production and flow direction alteration. Taking the orange box in Figure 12 as an example, the oil recovery efficiency considering the synergistic effect of increased liquid production and flow direction alteration after the change in well pattern is shown in Figure 14a. Considering the blue box in Figure 11, the oil recovery efficiency due to increased liquid production is shown in Figure 14b.

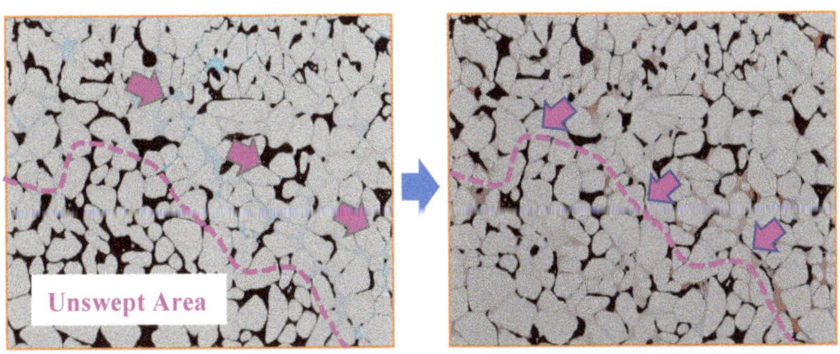

(**a**) Synergetic effect of liquid extraction and liquid flow diversion on oil displacement

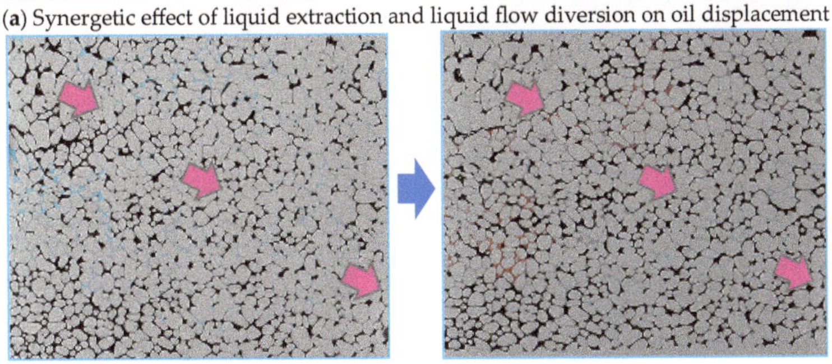

(**b**) Improvement in oil displacement by liquid extraction

Figure 14. Improvement in oil displacement effect by different mechanisms of well network transformation. The pink arrow represents the direction of fluid flow. The pink dashed line separates two regions with different fluid flow conditions.

Figure 14 shows that after the change in well pattern, the orange box region shifted from a non-mainstream area to a mainstream area. Due to the dual effect of increased liquid production and flow direction alteration, the previously unswept area was effectively swept, significantly improving oil recovery efficiency.

In contrast, the blue box region remained a mainstream area after the change in well pattern. Since the injection speed remained unchanged before and after the transformation, only the mechanism of increased liquid production contributed to the improvement in oil recovery efficiency. As seen in Figure 14b, even after transitioning to a two-injection and six-production pattern, this area did not complete the mixing process, indicating limited improvement. The oil recovery efficiency of the two regions before and after the change in well pattern is shown in Figure 15.

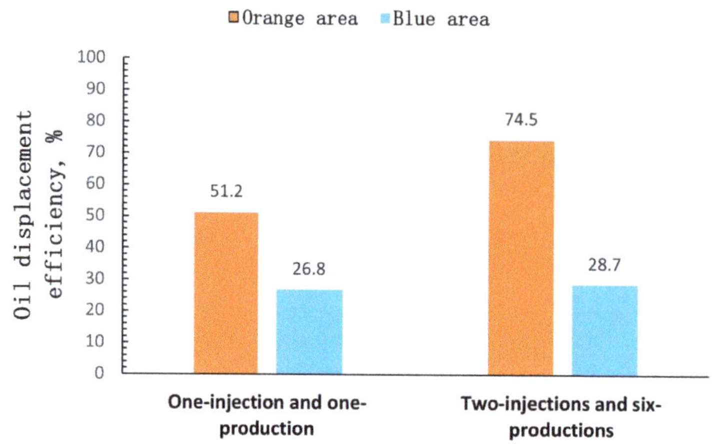

Figure 15. Comparative analysis of oil displacement efficiency in typical areas after well network transformation.

Figure 15 shows that the synergistic effect of increased liquid production and flow direction alteration in the orange region increased oil recovery efficiency from 51.2% to 74.5%. In the blue region, where only increased liquid production was applied, oil recovery efficiency increased from 26.8% to 28.7%. The synergistic effect of increased liquid production and flow direction alteration significantly enhanced oil recovery efficiency.

6. Conclusions and Insights

1. We proposed the use of a mixing method for tracing water flooding permeability channels and visualizing the evolution characteristics of dominant channels in ultra-high water cut reservoirs: targeted advancement in low water cut stages, lateral expansion in medium-to-high water cut stages, and contraction and solidification in ultra-high water cut stages.
2. During the ultra-high water cut period, dominant water flooding channels control the remaining oil distribution. The barrier effect divides the swept area into dominant and stagnant flow zones. The dominant flow zones exhibit near-single-phase flow, making the remaining oil highly dispersed and difficult to mobilize. This contributes to high water consumption in the late ultra-high water cut period. In contrast, the stagnant flow zones contain pores that do not participate in flow, resulting in relatively enriched remaining oil, representing important areas for potential recovery enhancement during the ultra-high water cut period.
3. Increased liquid production regulates the dominant water flooding zones through non-selective flow, while flow direction alteration reconstructs the water flooding flow field. Flow direction alteration initially improves oil recovery efficiency significantly, but

the secondary formation of dominant water flooding channels limits the effectiveness over time.

4. Physical simulation experiments demonstrate that the synergistic regulation of water flooding permeability channels through increased liquid production and flow direction alteration can significantly expand the sweep and enhance oil recovery efficiency, achieving a recovery rate of up to 68.02% during the ultra-high water cut period. This indicates that significant improvements in recovery rates can still be achieved through coordinated adjustments in the late ultra-high water cut period.
5. The formation of high water consumption channels is an inevitable phenomenon for field practice, and EOR can be further enhanced by increasing water injection intensity and changing injection direction. However, there are two problems: On the one hand, the enhanced strength of injected water must be enough to overcome the effect of the seepage barrier at the boundary of the dominant seepage area in order to improve the EOR. On the other hand, changing injection direction can significantly improve oil recovery, but there is a risk of secondary formation of the dominant seepage area, so the flow direction of injected water must be adjusted dynamically. The method for improving oil recovery by enhancing injection strength and changing injection direction through well-pattern infilling is a promising technology.

Author Contributions: Conceptualization, C.Y.; Methodology, C.Y.; Validation, S.Z.; Formal analysis, S.W.; Investigation, M.Z. and S.Z.; Resources, S.W.; Data curation, W.C.; Writing—original draft, C.Y. and S.W.; Writing—review & editing, W.C.; Visualization, M.Z.; Supervision, S.W.; Project administration, S.W. All authors have read and agreed to the published version of the manuscript.

Funding: This research is supported by National Natural Science Foundation of China (52374051).

Data Availability Statement: The original contributions presented in the study are included in the article, further inquiries can be directed to the corresponding author.

Conflicts of Interest: Authors Chunlei Yu, Min Zhang and Shiming Zhang were employed by the company China Petroleum and Chemical Corporation Shengli Oilfield Branch. The remaining authors declare that the research was conducted in the absence of any commercial or financial relationships that could be construed as a potential conflict of interest.

References

1. Gautier, Y.; Noetinger, B. Preferential flow-paths detection for heterogeneous reservoirs using a new renormalization technique. *Transp. Porous Media* **1997**, *26*, 1–23. [CrossRef]
2. Tang, Y.B.; Li, M.; Bernabé, Y.; Zhao, J.Z. Viscous fingering and preferential flow paths in heterogeneous porous media. *J. Geophys. Res. Solid Earth* **2020**, *125*, e2019JB019306. [CrossRef]
3. Li, Y. Study on enhancing oil recovery of continental reservoir by water drive technology. *Acta Pet. Sin.* **2009**, *30*, 396–399.
4. Yuan, S.; Wang, Q. New progress and prospect of oilfields development technologies in China. *Pet. Explor. Dev.* **2018**, *45*, 657–668. [CrossRef]
5. Sun, H.; Yang, Y.; Wang, H. Distribu-tion characteristics of remaining oil inextra-high water cut reservoirs and new technologies for enhancing oil recovery. *J. China Univ. Pet.* **2023**, *47*, 90–102.
6. Jiang, H. Early-warning and differentiated adjustment methods for channeling in oil reservoirs at ultra-high water cut stage. *J. China Univ. Pet.* **2013**, *37*, 114–119.
7. de Holanda, R.W.; Gildin, E.; Jensen, J.L.; Lake, L.W.; Kabir, C.S. A state-of-the-art literature review on capacitance resistance models for reservoir characterization and performance forecasting. *Energies* **2018**, *11*, 3368. [CrossRef]
8. Kazem, M.H.; Hussein, M.A.; Adnan, M.S.; Alfarge, D.; Mansour, I.J. The performance of streamline simulation technique to mimic the waterflooding management process in oil reservoirs. *Fuel* **2023**, *348*, 128556. [CrossRef]
9. Feng, Q.; Shi, F.; Wang, S.; Li, C.; Liu, G.; Tao, L. Determination of reasonable bottom hole flow pressure in fluid extraction wells. *Pet. Geol. Recovery Effic.* **2011**, *18*, 74–76.
10. Gentil, P.H. The Use of Multi-Linear Regression Models in Patterned Waterfloods: Physical Meaning of the Regression Coefficients. Doctoral Dissertation, The University of Texas, Austin, TX, USA, 2005.
11. Yousef, A.A.; Gentil, P.H.; Jensen, J.L.; Lake, L.W. A capacitance model to infer interwell connectivity from production- and injection-rate fluctuations. *SPE Reserv. Eval. Eng.* **2006**, *9*, 630–646. [CrossRef]
12. Parra, J.E.; Samaniego, V.F.; Lake, L.W. Application of the producer-based capacitance resistance model to undersaturated oil reservoirs in primary recovery. *SPE J.* **2023**, *28*, 2256–2273. [CrossRef]

13. Guo, Z.; Reynolds, A.C.; Zhao, H. Waterflooding optimization with the INSIM-FT data-driven model. *Comput. Geosci.* **2018**, *22*, 745–761. [CrossRef]
14. Guo, Z.; Reynolds, A.C. INSIM-FT in three-dimensions with gravity. *J. Comput. Phys.* **2019**, *380*, 143–169. [CrossRef]
15. Lie, K.A.; Krogstad, S. Data-driven modelling with coarse-grid network models. *Comput. Geosci.* **2023**, *28*, 273–287. [CrossRef]
16. Wang, D.; Li, Y.; Chen, B.; Hu, Y.; Li, B.; Gao, D.; Fu, B. Ensemble-based optimization of interwell connectivity in heterogeneous waterflooding reservoirs. *J. Nat. Gas Sci. Eng.* **2017**, *38*, 245–256. [CrossRef]
17. Li, Y.; Alpak, F.O.; Jain, V.; Lu, R.; Onur, M. History matching and forecasting production rate and bottomhole pressure data using an enhanced physics-based data-driven simulator. *SPE Reserv. Eval. Eng.* **2023**, *26*, 957–974. [CrossRef]
18. Li, Y.; Onur, M. INSIM-BHP: A physics-based data-driven reservoir model for history matching and forecasting with bottomhole pressure and production rate data under waterflooding. *J. Comput. Phys.* **2023**, *473*, 111714. [CrossRef]
19. Kalaydjian, O.V.R. Study of Wettability and Spreading Impact in Three-Phase Gas Injection by Cryo-Scanning Electron Microscopy. *J. Pet. Sci. Eng.* **1998**, *20*, 189–202.
20. Liu, J. Analysis of reservoir physical properties and influencing factors of high water cut sandstone reservoir in Gaoyou sag. *Pet. Geol. Eng.* **2019**, *33*, 53–57.
21. Liu, Y.; Sun, B.; Yu, Y. Fuzzy identification and quantative calculation method for big pore throat. *Oil Drill. Prod. Technol.* **2003**, *25*, 54–59.
22. Zhong, D.; Zhu, X.; Wu, S.; Jin, S.; Jia, D.; Zhao, Y. Characteristics and controlling factors of high capacity channels of reservoirs at high water cut stage: A case from Block Hu 12 of Huzhuangji Oilfield. *Pet. Explor. Dev.* **2007**, *34*, 207–211.
23. Yao, F.; Yao, T.; Li, J. Characteristies and influencing factors of reservoir wettability reversal. *Pet. Geol. Recover.* **2007**, *14*, 76–78.
24. Huang, Y.; Tang, Y.; Li, L. Changes in reservoirs with mid-low permeability after water flooding and analysis of influencing factors. *J. Oil Gas Technol.* **2013**, *35*, 143–147.
25. Wang, D. Some thoughts about potential of oil recovery efficiency and development model transition in Shengli district. *Pet. Geol. Recovery Effic.* **2014**, *21*, 1–4.
26. Zhang, C.; Zheng, C.; Xiao, W. Research on the influencing factors and functional mechanism of enhanced liquid result in ultra-high water cut stage-case of 7~4-8~1 unit of Es_2 in block No.2 of Shengtuo oilfield. *Pet. Geol. Recovery Effic.* **2013**, *20*, 88–91.
27. Bai, M.; Zhang, Z.; Li, Y.; Zhang, C. Study on Precise Flooding Direction Adjustment of Sandstone Oilfields in the Late Stage with High Water Cut. *Spec. Oil Gas Reserv.* **2017**, *24*, 86–88.
28. Cao, T.; Gao, D.; Li, Z.; Wang, T.; Jiang, F. Characteristics of Dominant Flowing Channels and Throat Volume of Multi-Layer Sandstone Reservoirs in Liuzan Oilfield. *Xinjiang Pet. Geol.* **2015**, *45*, 53–57.
29. Yuan, Q.; Zhu, L.; Lu, H.; Zheng, X. Development characteristics and main tackled EOR research direction for the water flooded oilfield at the late stage. *Pet. Geol. Oilfield Dev. Daqin* **2019**, *38*, 34–40.
30. Wang, Z.; Zhang, G.; Jin, Y.; Pei, H.; Shi, S.; Jiang, P. Classification criteria and volume calculation method for different graded waterflooding zones of uncompartmentalized oilfields at ultra-high water cut stage. *Pet. Geol. Recovery Effic.* **2022**, *29*, 75–82.
31. Lin, Y. Forming Mechanism of the Preferential Seepage Channel for the Reservoir at the Late Stage of the High Water cut. *Pet. Geol. Oilfield Dev. Daqing* **2018**, *37*, 33–37.
32. Wang, M.; Shi, C.; Zhu, W.; Ding, L. Identification and accurate description of preponderance flow path. *Pet. Geol. Recovery Effic.* **2016**, *23*, 79–84.
33. Zhang, M.; Wang, Z.; Yang, Y.; Sun, Y.; Sun, B.; Zhang, L. Analysis on macroscopic displacement effect of hot water flooding on heavy oil by "Cut-Flatten Image" method. *Pet. Geol. Recovery Effic.* **2016**, *23*, 85–90.

Disclaimer/Publisher's Note: The statements, opinions and data contained in all publications are solely those of the individual author(s) and contributor(s) and not of MDPI and/or the editor(s). MDPI and/or the editor(s) disclaim responsibility for any injury to people or property resulting from any ideas, methods, instructions or products referred to in the content.

energies

MDPI

Article

A Study on the Thermal Physical Property Changes in Formation Rocks during Rapid Preheating of SAGD

Jie Tian [1,*], Shiwen Huang [2], Mingda Dong [1], Wende Yan [1,*] and Zhilin Qi [1,*]

[1] School of Petroleum and Natural Gas Engineering, Chongqing University of Science & Technology, Chongqing 401331, China; 2019063@cqust.edu.cn
[2] Experimental Testing Technology Center of Northwest Oilfield Branch of Sinopec, Urumqi 830011, China; huangshw.xbsj@sinopec.com
* Correspondence: tianjie@cqust.edu.cn (J.T.); 2012020@cqust.edu.cn (W.Y.); 2008008@cqust.edu.cn (Z.Q.)

Abstract: The incorporation and application of SAGD rapid preheating technology effectively solve the problem of the long preheating cycle in the SAGD steam cycle. The thermal properties of reservoir rocks are an important factor affecting the heat transfer law governing their formation during the rapid preheating process of SAGD. During the rapid preheating process of SAGD, the expansion of the reservoir and the steam cycle process will cause changes in the pore permeability, oil-water saturation, and temperature of the reservoir rocks, which will inevitably lead to differences in the changes that occur in the thermal properties of the reservoir rocks, compared to those under the influence of a single factor. In this study, experiments were conducted to determine the thermal properties of reservoir rocks under the combined influence of pore permeability, oil-water saturation, and temperature, quantitatively characterizing the changes in the thermal properties of reservoir rocks. Using the orthogonal method to design and carry out experiments for determining the thermal properties of reservoir rocks, the main factors affecting the thermal properties of reservoir rocks and the significance of each factor's impact on the thermal properties of reservoir rocks were determined through intuitive analysis and variance analysis of the experimental results. Finally, a regression equation that can characterize changes in the thermal properties of reservoir rocks under the influence of multiple factors was obtained through multiple nonlinear regressions of the experimental results.

Keywords: SAGD; rapid preheating; thermal properties

1. Introduction

Cyclic steam stimulation and steam circulation are commonly used as SAGD preheating methods. Compared to cyclic steam stimulation preheating, steam circulation preheating heats the formation relatively uniformly and has better connectivity, so steam circulation preheating is the most widely used preheating technology. Although the preheating effect of the steam cycle is better, its long preheating time not only leads to huge steam consumption but also produces a large amount of difficult-to-treat oil residue during the circulation process. These problems increase the development cost of SAGD technology to some extent and have a negative impact on the improvement of economic benefits. In order to solve the problem of the long preheating time, the BitCan Company in Canada proposed the use of SAGD rapid preheating technology in 2015 and it has been adopted rapidly both domestically and internationally. The SAGD rapid preheating technology utilizes the stress dilatation principle in rock mechanics to expand the rock of the reservoir near the wellbore before preheating, causing changes in the rock particle structure of the reservoir, forming a dilatation area with roughly the same increase in permeability and porosity. This allows the steam to develop thermal connectivity between the two wells more quickly, achieving the goal of shortening the preheating time. During the rapid preheating process of SAGD, parameters such as reservoir porosity, oil-water saturation, and temperature will change, all of which will cause changes in the thermal properties of the reservoir, thereby affecting the application of the heat transfer law in the reservoir. The thermal properties of reservoir rocks mainly include thermal conductivity, specific heat capacity, and the thermal diffusion coefficient. There are many factors that can affect the thermal properties of reservoir rocks, including their petrological characteristics, porosity, oil-water saturation, pressure, and temperature. Xu Zhenzhang from China [1] systematically elaborated on the factors affecting the thermal properties of reservoir rocks in 1992 and analyzed the mechanisms of changes in reservoir rock thermal properties under the influence of various factors based on research data at that time. However, due to the limited testing technology and instrument conditions of early rock thermal properties testing, testing could only be conducted at room temperature and pressure, which cannot effectively simulate geological conditions. Therefore, the accuracy of rock thermal property testing results is poor [2]. In recent years, there has been significant development in testing technology and instruments, with significantly improved accuracy of test results and the ability to effectively simulate various temperature and pressure conditions for determining rock thermal properties.

Considering the influence of temperature on the thermal properties of rocks, Sun et al. [3] measured the thermal properties of dry sandstone under different temperature conditions and described the changes in thermal properties of the sandstone under conditions of 25 °C to 900 °C in four stages, based on the measurement results. The thermal conductivity of the rock showed an overall decreasing trend, while the specific heat capacity was proportional to temperature before 200 °C, fluctuating between 200 °C and 400 °C, and was inversely proportional to temperature after 600 °C. Abdulagaov [4] and Emirov [5] also reached similar research conclusions. Geng et al. [6] investigated the effect of temperature on the thermal diffusion coefficient through experimental measurements and found that the thermal diffusion coefficient of sandstone is inversely proportional to temperature and tends to stabilize when the temperature reaches 600 °C or above. As a result of such analysis, it is believed that the decrease in thermal diffusion coefficient from 25 °C to 300 °C is mainly due to the escape of attached water, bound water, and structural water. Between 300 and 600 °C, the thermal response of minerals in sandstone increases the development of microcracks and weakens the thermal diffusion coefficient of sandstone. Unlike in previous research, Liu et al. [7] conducted experimental and modeling studies on heat transfer in sandstone under low-temperature conditions, and conducted thermal property tests on saturated water, saturated oil, and dry sandstone under low-temperature conditions (−196.13–19.85 °C). The test results show that the thermal conductivity of dry sandstone under low-temperature conditions increases with an increase in temperature, which is different from the situation where the thermal conductivity decreases with an increase in temperature under high-temperature conditions. Analysis suggests that this is mainly related to the strong phase transition, which absorbs a large amount of latent heat.

Considering the influence of oil-water saturation on the thermal properties of rocks, Guo Yeping et al. [8] measured the thermal conductivity of sandstone under different temperature and water-content conditions. The analysis showed that the thermal conductivity of sandstone is inversely proportional to temperature, and the thermal conductivity of sandstone in a saturated state will undergo a sudden change near 0 °C due to the influence of the water phase change. However, under a constant temperature, the thermal conductivity of sandstone is directly proportional to the water content, and the growth rate of thermal conductivity is inversely proportional to the ambient temperature. Hu Rong et al. [9] conducted a measurement study on the thermal properties of rocks in the Chunguang Oilfield, analyzing the effects of oil-water saturation and rock density on thermal properties. The study found that the thermal conductivity and specific heat capacity of rocks are directly proportional to their water saturation and density, and are inversely proportional to their oil saturation. In addition, by regressing the results of thermal property testing, a regression relationship equation was obtained for calculating rock thermal properties through the difference in logging acoustic time. Song Xiaoqing et al. [10] conducted a study on the thermal properties of the main rocks in Guizhou, and the results showed that the thermal conductivity of rocks under saturated water conditions increased by 2–17% compared to those under dry conditions, the thermal diffusion coefficient increased by 1–16%, and the specific heat capacity decreased by 3.08–21.79%. Analysis suggests that in addition to the petrological characteristics, the water content of rocks is the main factor affecting their thermal properties. Zhen Zuolin et al. [11] conducted experiments to determine the thermal properties of underground transportation surrounding certain rocks in Lanzhou. The experimental results showed that the thermal conductivity of the surrounding soil samples increased linearly with the increase in water content; the volumetric specific heat capacity decreased first and then increased with the increase in water content, and the thermal diffusion coefficient increased first and then slowly decreased with the increase in water content.

In terms of the impact of changes in pore permeability on the thermal properties of rocks, Scharli et al. [12] and Sayed et al. [13] conducted studies on the influence of rock porosity on rock thermal properties in 1984 and 2011, respectively. The results showed that the porosity of rocks was inversely proportional to their thermal conductivity, but this result was only based on data analysis and did not provide empirical evidence. In 2020, Zhu et al. [14] used scanning electron microscopy to obtain images of rock samples, analyzed the microstructure of different rock types, and studied the relationship between thermal conductivity and porosity using eight different thermal conductivity porosity models. The results indicate that the thermal conductivity decreases with an increase in porosity, confirming the previous research findings.

Given the impact of changes in pore permeability on the thermal properties of rocks, studies by Scharli et al. [12] in 1984 and Sayed et al. [13] in 2011 investigated the relationship between rock porosity and thermal properties. These findings indicated that the porosity of rocks is inversely proportional to their thermal conductivity, though these results were solely based on data analysis and were not empirical. In 2020, Zhu et al. [14] utilized scanning electron microscopy to capture images of rock samples and examined the microstructure of various rock types. They further examined the correlation between thermal conductivity and porosity using eight distinct thermal conductivity-porosity models. Their results suggest that thermal conductivity decreases as porosity increases, corroborating the previous research findings.

In predicting the thermal properties of rocks, various models have been proposed [15–21], all of which are obtained through regression methods based on a large amount of experimental data, and these models only consider the influence of single factors for thermal property prediction. However, there are many factors that affect the thermal properties of reservoir rocks, and the direction of their influence is also different. Therefore, many prediction models that only consider the influence of single factors have significant limitations.

Overall, there is currently a wealth of research on changes in reservoir thermal properties, and scholars have conducted quantitative or qualitative studies on the various factors that affect reservoir rock thermal properties. They have summarized the mechanisms and predictive models of reservoir thermal property changes under the influence of corresponding factors. However, the shortcomings are that currently, most research conclusions are obtained through experimental analysis under the control of single-factor variables, and there is no research on the changes in reservoir thermal properties under the joint influence of multiple factors. For the rapid preheating process of SAGD in the F reservoir studied in this study, factors affecting the thermal properties of reservoir rocks include increased local reservoir porosity and permeability after expansion, as well as changes in oil-water saturation and temperature that are caused by steam entering the formation. From previous research, it can be seen that the influence of these factors on thermal properties, especially thermal conductivity, is not consistent. Therefore, the changes in rock thermal properties under the combined action of multiple factors is a problem worthy of in-depth research, and it is also the basis for conducting subsequent research on heat transfer laws in reservoirs.

2. Apparatus and Procedures

2.1. Materials

The oil for the experiments (Figure 1) was taken from reservoir F; the viscosity of crude oil was 39,952 mPa·s under reservoir temperature (22 °C) conditions. The rock cores used in the experiment are shown in Figure 2.

Figure 1. Crude oil used in the experiment.

Figure 2. Cores used in the experiment.

2.2. Apparatus and Procedures

2.2.1. Oil-Water Core Saturation

Before conducting thermal property measurements, it is necessary to fully saturate the core sample with oil and water according to the experimental design requirements and slice it into slices. The core saturation device used in the experiment includes a formation fluid saturation system (Figure 3) and an oil-water saturation device (Figure 4).

The formation fluid saturation system is mainly used in experiments to saturate the formation water of artificial rock cores. The specific steps are as follows:

(1) Prepare a sufficient amount of formation water according to the experimental requirements and fill a water tank with it;
(2) After placing the rock core into the rock ventricle, close the rock ventricle and turn on the rock ventricle vacuum pump, continuously vacuuming it for 48 h;
(3) Turn off the vacuum pump and open the valve between the formation water tank and the rock core chamber;

(4) Slowly apply pressure to the rock ventricle using a hand pump;
(5) Stop pressurization when the pressure of the rock ventricle remains unchanged for 12 h;
(6) Reduce the pressure of the rock ventricle to 0 MPa, open the vent valve to drain excess formation water, remove the rock core, inspect and clean the equipment, and prepare for the next experiment.

Figure 3. Formation fluid saturation system.

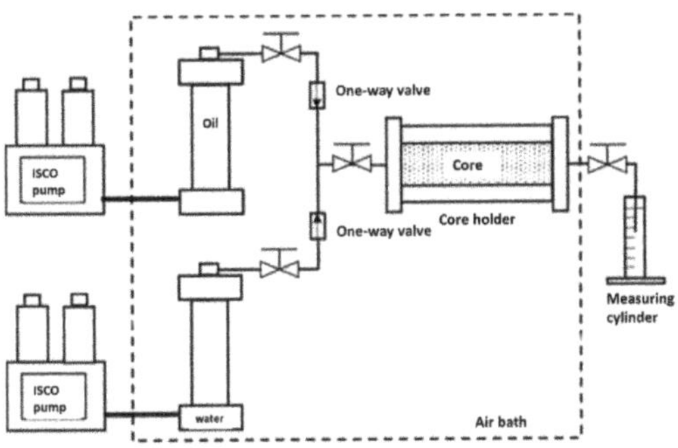

Figure 4. Steady-state oil-water saturation device.

Due to the strict requirements for core oil-water saturation in thermal property testing experiments, conventional oil-water saturation methods cannot accurately control the oil-water saturation of artificial cores. Therefore, by referring to the steady-state method for measuring the oil-water permeability of rock cores, the oil-water saturation of the saturated water rock cores was controlled. The experimental setup flowchart is shown in Figure 4. Two constant flow pumps are used to pump oil and water into the rock core. During this process, the proportion of the oil and water injection rate remains constant. When the pump pressure is stable and the produced oil-water ratio remains consistent with the injected oil-water ratio for 20 min, it can be considered that the oil-water ratio in the rock core pores is the same as the injected oil-water ratio, which achieves the goal of controlling the oil-water saturation of the core. The specific steps are as follows:

(1) Connect up the equipment and open the constant temperature box at least 30 min before the experiment, then adjust the temperature of the constant temperature box to 80 °C;
(2) Place the saturated formation water core into the core gripper, then connect the core gripper to the device according to the flowchart;

(3) Close the outlet valve of the rock core gripper, open the ISCO pump to inject formation water into the rock core, and check the sealing;
(4) After confirming the sealing of the device, open the outlet valve of the gripper, and after the water stabilizes, close the ISCO pump and all valves;
(5) Turn on two ISCO pumps, adjust to the specified flow rate according to experimental requirements, open all valves, and saturate the core with oil and water simultaneously;
(6) After the oil and water at the outlet end of the core gripper are discharged for 5 min, use a measuring cylinder at the outlet end to collect the produced liquid in time intervals (10–20 min each time) and observe the pressure changes of the ISCO pump. When the pump pressure is basically stable and the oil-water ratio of the produced liquid is the same as the injected oil-water ratio, close the ISCO pump and all valves;
(7) Close the constant-temperature box, lower the temperature of the gripper to room temperature, then open the gripper and remove the rock core;
(8) Clean the device and prepare for the next experiment.

2.2.2. Thermophysical Property Determination

The experimental device for measuring thermal properties adopts the LFA467 laser thermal conductivity instrument produced by the NETZSCH company in Selby, Germany, as shown in Figure 5.

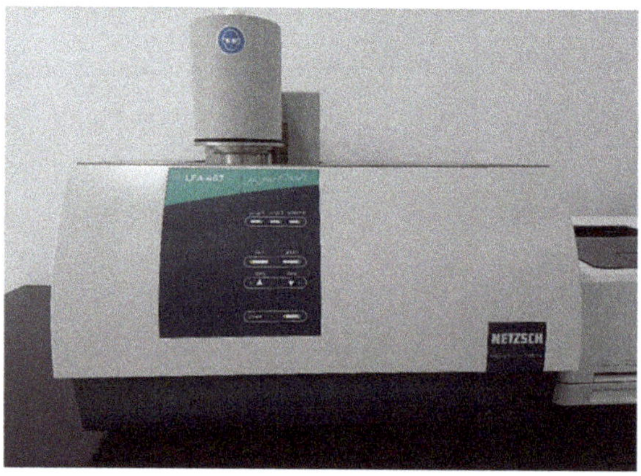

Figure 5. LFA467 laser thermal conductivity meter.

The specific testing steps are as follows:

(1) Turn on the instrument power and wait for about 10 s before the "unlock" light turns on. Simultaneously press the "close + safety" button until the "close" light remains on and the instrument is ready;
(2) Turn on the computer and water bath, set the water bath temperature to 2 °C above room temperature;
(3) Add liquid nitrogen to the infrared detector and stabilize it for 30 min before testing begins;
(4) Use nitrogen as the blowing gas and set the output pressure to 0.05 MPa;
(5) Cut the saturated oil and water core into rock sample slices according to the experimental requirements (as shown in Figure 6), and spray graphite on the upper and lower sides of the slices;
(6) Simultaneously press the "close + safety" button to open the injection port, place the sample into the instrument, and prepare for testing (as shown in Figure 7);
(7) Open the measurement software, set the measurement parameters, and start the test (as shown in Figure 8);

(8) After the test is completed, open the analysis software to analyze the test results and output the test report;
(9) When the temperature of the sample tray cools to below 100 °C, open the furnace and take out the sample.

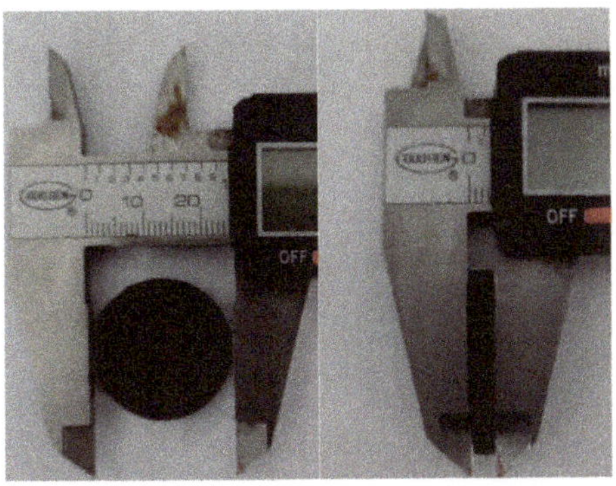

Figure 6. Rock samples after slicing.

Figure 7. Samples to be tested.

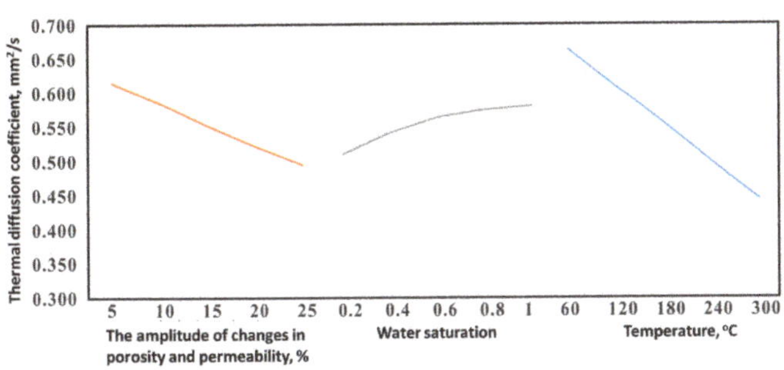

Figure 8. Effect curve of thermal diffusivity.

During the experiment, attention should be paid to protecting the eyes and other parts of the human body from ultra-low-temperature burns when adding liquid nitrogen; similarly, be careful of high-temperature components when opening the furnace body to avoid high-temperature burns.

2.2.3. Experimental Plan and Parameter Design

In order to study the changes in reservoir thermal properties under the influence of multiple factors through experiments, based on survey data and laboratory conditions, three influencing factors were selected: pore permeability change amplitude, water saturation, and temperature (Table 1). Five levels were selected for each factor, and a six-factor five-level orthogonal design table (three empty columns) was used to carry out a three-factor five-level orthogonal experimental design (Table 2).

Table 1. Orthogonal experimental factor level table.

Level	Factors		
	A Amplitude of Changes in Porosity and Permeability, %	B Water Saturation	C Temperature, °C
1	5	0.2	60
2	10	0.4	120
3	15	0.6	180
4	20	0.8	240
5	25	1	300

Table 2. Experimental scheme design table of thermal property measurement.

No.	Amplitude of Changes in Porosity and Permeability, %	Water Saturation	Temperature, °C
1	5	0.2	60
2	5	0.4	120
3	5	0.6	180
4	5	0.8	240
5	5	1	300
6	10	0.2	120
7	10	0.4	180
8	10	0.6	240
9	10	0.8	300
10	10	1	60
11	15	0.2	180
12	15	0.4	240
13	15	0.6	300
14	15	0.8	60
15	15	1	120
16	20	0.2	240
17	20	0.4	300
18	20	0.6	60
19	20	0.8	120
20	20	1	180
21	25	0.2	300
22	25	0.4	60
23	25	0.6	120
24	25	0.8	180
25	25	1	240

Due to limitations in the experimental conditions, it is necessary to manufacture artificial rock cores as thermal property measurement samples, based on the pore permeability conditions, when conducting thermal property measurement experiments. The parameters of the core samples are shown in Table 3.

Table 3. Experimental scheme design table of thermal property measurement.

No.	Permeability, D	Porosity
1	0.885	0.3175
2	0.885	0.3175
3	0.882	0.312
4	0.882	0.312
5	0.899	0.3559
6	0.916	0.3605
7	0.916	0.3605
8	0.924	0.327
9	0.924	0.327
10	0.924	0.312
11	0.965	0.346
12	0.965	0.346
13	0.965	0.331
14	0.965	0.331
15	0.975	0.3632
16	1.008	0.3637
17	1.008	0.3637
18	1.009	0.3650
19	1.009	0.3650
20	1.015	0.378
21	1.050	0.358
22	1.050	0.358
23	1.050	0.364
24	1.050	0.364
25	1.050	0.3507

3. Results and Discussion

The experimental results for the samples' thermal properties obtained from the LFA thermal conductivity tester are shown in Table 4.

Table 4. Experimental scheme design table of thermal property measurement.

No.	Amplitude of Changes in Porosity and Permeability, %	Water Saturation	Temperature °C	Thermal Diffusion Coefficient, mm^2/s	Thermal Conductivity Coefficient, W/(m·K)	Specific Heat, J/(g·K)
1	5	0.2	60	0.751	1.422	1.790
2	5	0.4	120	0.690	1.360	1.820
3	5	0.6	180	0.622	1.330	1.829
4	5	0.8	240	0.540	1.331	1.811
5	5	1	300	0.469	1.360	1.770
6	10	0.2	120	0.601	1.199	1.800
7	10	0.4	180	0.570	1.180	1.830
8	10	0.6	240	0.544	1.180	1.820
9	10	0.8	300	0.500	1.220	1.801
10	10	1	60	0.711	1.301	1.570
11	15	0.2	180	0.480	0.990	1.833
12	15	0.4	240	0.480	1.000	1.850
13	15	0.6	300	0.492	1.050	1.840
14	15	0.8	60	0.690	1.270	1.590
15	15	1	120	0.620	1.190	1.572
16	20	0.2	240	0.387	0.793	1.880
17	20	0.4	300	0.430	0.840	1.890
18	20	0.6	60	0.630	1.220	1.611
19	20	0.8	120	0.602	1.141	1.610
20	20	1	180	0.560	1.100	1.590
21	25	0.2	300	0.330	0.600	1.910
22	25	0.4	60	0.533	1.120	1.647
23	25	0.6	120	0.530	1.050	1.673
24	25	0.8	180	0.540	1.020	1.660
25	25	1	240	0.540	1.149	1.630

3.1. Intuitive Analysis and Analysis of Variance

3.1.1. Intuitive Analysis

Based on the orthogonal experimental factor level table and experimental results, a visual analysis is conducted on the test results for the thermal diffusion coefficient, thermal conductivity coefficient, and specific heat. The vacant columns (D, E, F) are used as error columns and have no practical significance. In the intuitive analysis table (Tables 5–7), Ki is the sum of the experimental values of the level i of the corresponding factors in the column; ki is the average experimental value of the level i of the corresponding factors in the column, ki = Ki/number of levels; R is the range of the mean values at each level.

Table 5. Thermal diffusivity experimental results: intuitive analysis table.

No.	Factor						Thermal Diffusion Coefficient, mm^2/s
	A	B	C	D	E	F	
1	1	1	1	1	1	1	0.75
2	1	2	2	2	2	2	0.69
3	1	3	3	3	3	3	0.62
4	1	4	4	4	4	4	0.54
5	1	5	5	5	5	5	0.47
6	2	1	2	3	4	5	0.6
7	2	2	3	4	5	1	0.57
8	2	3	4	5	1	2	0.54
9	2	4	5	1	2	3	0.5
10	2	5	1	2	3	4	0.71
11	3	1	3	5	2	4	0.48
12	3	2	4	1	3	5	0.48
13	3	3	5	2	4	1	0.49
14	3	4	1	3	5	2	0.69
15	3	5	2	4	1	3	0.62
16	4	1	4	2	5	3	0.39
17	4	2	5	3	1	4	0.43
18	4	3	1	4	2	5	0.63
19	4	4	2	5	3	1	0.6
20	4	5	3	1	4	2	0.56
21	5	1	5	4	3	2	0.33
22	5	2	1	5	4	3	0.53
23	5	3	2	1	5	4	0.53
24	5	4	3	2	1	5	0.54
25	5	5	4	3	2	1	0.54
K_1	3.072	2.549	3.315	2.821	2.885	2.955	
K_2	2.926	2.703	3.043	2.820	2.840	2.814	
K_3	2.762	2.818	2.772	2.883	2.745	2.662	
K_4	2.609	2.872	2.491	2.690	2.726	2.691	
K_5	2.473	2.900	2.221	2.628	2.646	2.720	
k_1	0.614	0.510	0.663	0.564	0.577	0.591	
k_2	0.585	0.541	0.609	0.564	0.568	0.563	
k_3	0.552	0.564	0.554	0.577	0.549	0.532	
k_4	0.522	0.574	0.498	0.538	0.545	0.538	
k_5	0.495	0.580	0.444	0.526	0.529	0.544	
R	0.120	0.070	0.219	0.051	0.048	0.059	

From the data in the table above, it can be seen that the range values R_A, R_B, and R_C, corresponding to the amplitude of pore permeability change (A), water saturation (B), and temperature (C), are arranged in descending order: $R_C > R_A > R_B$. This indicates that the change in temperature (C) has the greatest impact on the thermal diffusion coefficient of the reservoir among the three factors studied in this article and is the main factor affecting changes in the thermal diffusion coefficient. Therefore, the order of the degree of influence on the thermal diffusion coefficient of the reservoir among the three factors is: temperature (C) > amplitude of pore permeability change (A) > water saturation (B). Similarly, the range of factors that affect the thermal conductivity of reservoirs is ranked as follows: $R_A > R_C > R_B$, and the amplitude of pore permeability change (A) is the main factor affecting the thermal conductivity of reservoir rocks. The order of magnitude of the impact of the three factors on the thermal conductivity of reservoir rocks is: pore permeability change amplitude (A) > temperature (C) > water saturation (B). The range order of factors affecting the specific heat of reservoir rocks is $R_B > R_C > R_A$, and the water saturation (B) is the main factor affecting the specific heat of reservoir rocks. The order of the degree of influence of the three factors on the specific heat of reservoir rocks is: water saturation (B) > temperature (C) > pore permeability change amplitude (A).

Table 6. Thermal conductivity experimental results: intuitive analysis table.

No.	Factor						Thermal Conductivity Coefficient, W/(m·K)
	A	B	C	D	E	F	
1	1	1	1	1	1	1	1.42
2	1	2	2	2	2	2	1.36
3	1	3	3	3	3	3	1.33
4	1	4	4	4	4	4	1.33
5	1	5	5	5	5	5	1.36
6	2	1	2	3	4	5	1.2
7	2	2	3	4	5	1	1.18
8	2	3	4	5	1	2	1.18
9	2	4	5	1	2	3	1.22
10	2	5	1	2	3	4	1.3
11	3	1	3	5	2	4	0.99
12	3	2	4	1	3	5	1
13	3	3	5	2	4	1	1.05
14	3	4	1	3	5	2	1.27
15	3	5	2	4	1	3	1.19
16	4	1	4	2	5	3	0.79
17	4	2	5	3	1	4	0.84
18	4	3	1	4	2	5	1.22
19	4	4	2	5	3	1	1.14
20	4	5	3	1	4	2	1.1
21	5	1	5	1	3	2	0.6
22	5	2	1	5	4	3	1.12
23	5	3	2	1	5	4	1.05
24	5	4	3	2	1	5	1.02
25	5	5	4	3	2	1	1.15
K_1	6.803	5.004	6.333	5.792	5.652	5.942	
K_2	6.080	5.500	5.940	5.524	5.939	5.510	
K_3	5.500	5.830	5.620	5.788	5.372	5.653	
K_4	5.094	5.982	5.453	5.521	5.800	5.512	
K_5	4.939	6.100	5.070	5.791	5.653	5.799	
k_1	1.361	1.001	1.267	1.158	1.130	1.188	
k_2	1.216	1.100	1.188	1.105	1.188	1.102	
k_3	1.100	1.166	1.124	1.158	1.074	1.131	
k_4	1.019	1.196	1.091	1.104	1.160	1.102	
k_5	0.988	1.220	1.014	1.158	1.131	1.160	
R	0.373	0.219	0.253	0.054	0.113	0.086	

Table 7. Visual analysis table of specific heat tests: experimental results.

No.	Factor						Specific Heat, J/(g·K)
	A	B	C	D	E	F	
1	1	1	1	1	1	1	1.79
2	1	2	2	2	2	2	1.82
3	1	3	3	3	3	3	1.83
4	1	4	4	4	4	4	1.81
5	1	5	5	5	5	5	1.77
6	2	1	2	3	4	5	1.8
7	2	2	3	4	5	1	1.83
8	2	3	4	5	1	2	1.82
9	2	4	5	1	2	3	1.8
10	2	5	1	2	3	4	1.57
11	3	1	3	5	2	4	1.83
12	3	2	4	1	3	5	1.85
13	3	3	5	2	4	1	1.84
14	3	4	1	3	5	2	1.59
15	3	5	2	4	1	3	1.57
16	4	1	4	2	5	3	1.88
17	4	2	5	3	1	4	1.89
18	4	3	1	4	2	5	1.61
19	4	4	2	5	3	1	1.61
20	4	5	3	1	4	2	1.59
21	5	1	5	4	3	2	1.91
22	5	2	1	5	4	3	1.65
23	5	3	2	1	5	4	1.67
24	5	4	3	2	1	5	1.66
25	5	5	4	3	2	1	1.63
K_1	9.020	9.213	8.208	8.704	8.732	8.700	
K_2	8.821	9.037	8.475	8.770	8.695	8.730	
K_3	8.685	8.773	8.742	8.739	8.769	8.729	
K_4	8.581	8.472	8.991	8.734	8.688	8.777	
K_5	8.520	8.132	9.211	8.680	8.743	8.691	
k_1	1.804	1.6264	1.642	1.741	1.746	1.740	
k_2	1.764	1.6944	1.694	1.754	1.738	1.746	
k_3	1.736	1.7546	1.748	1.748	1.754	1.746	
k_4	1.716	1.8074	1.798	1.746	1.738	1.754	
k_5	1.704	1.8426	1.842	1.736	1.748	1.738	
R	0.100	0.216	0.200	0.018	0.016	0.016	

We can map the effect curves of each indicator based on the data in the table as follows.

According to Figures 8–10, it can be seen that within the parameter range studied in this article, the thermal diffusion coefficient is negatively correlated with the amplitude of pore permeability changes, positively correlated with water saturation, and negatively correlated with temperature. The thermal conductivity is negatively correlated with the amplitude of pore permeability changes, positively correlated with water saturation, and negatively correlated with temperature. The specific heat is negatively correlated with the amplitude of pore permeability changes, positively correlated with water saturation, and positively correlated with temperature. The variation trends of the thermal diffusion coefficient, thermal conductivity coefficient, and specific heat obtained from orthogonal experiments with respect to each factor are consistent with the research results for single-factor influences in References [6–8], indicating that the results of this study are consistent with those in previous studies.

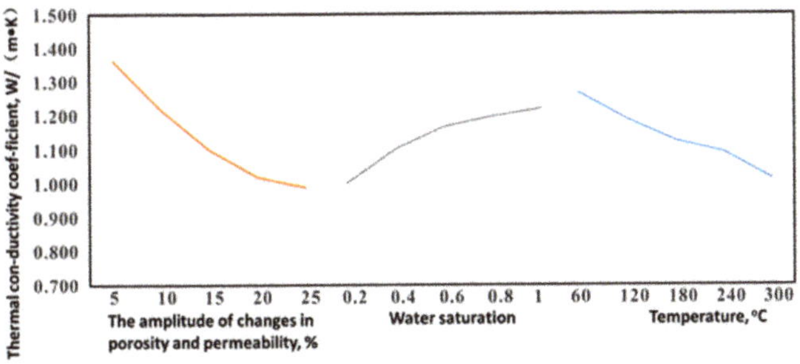

Figure 9. Effect curve of thermal conductivity.

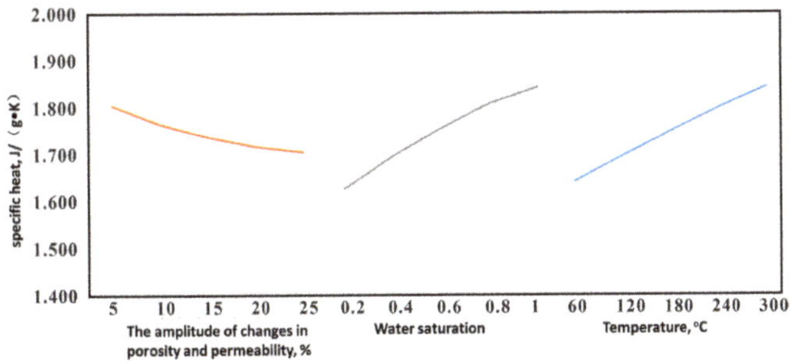

Figure 10. Effect curve of specific heat.

3.1.2. Variance Analysis

The main influencing factors of each indicator and the changing trends of the three indicators with each factor were determined through intuitive analysis in the previous section. However, intuitive analysis can only determine the magnitude of the impact of different factors on the indicators and cannot quantify the degree of impact. Therefore, an analysis of variance was chosen to determine the significance of the impact of each factor on different indicators. For this, we calculate the sum of squared deviations, degrees of freedom, and mean square of each factor that affects the thermal diffusion coefficient, thermal conductivity coefficient, and specific heat, based on the intuitive analysis table's F-value. The F-critical value and *p*-value were used for the analysis of variance, and the calculation results are shown in Tables 8–10 below.

Table 8. Thermal diffusion coefficient variance analysis table.

Factor	Sum of Squared Deviations	Degree of Freedom	Mean Square	F-Value	$F_{0.05}$	*p*-Value	Significance
A	0.0459	4	0.0115	5.0295	3.260	0.0129	Significant
B	0.0166	4	0.0041	1.8156	3.260	0.1907	Not significant
C	0.1502	4	0.0375	16.4378	3.260	0.0001	Significant
Error	0.0274	12					

Table 9. Thermal conductivity variance analysis table.

Factor	Sum of Squared Deviations	Degree of Freedom	Mean Square	F-Value	$F_{0.05}$	*p*-Value	Significance
A	0.4692	4	0.1173	17.3986	3.260	0.0001	Significant
B	0.1559	4	0.0390	5.7806	3.260	0.0079	Significant
C	0.1842	4	0.0461	6.8319	3.260	0.0042	Significant
Error	0.0809	12					

Table 10. Specific heat variance analysis table.

Factor	Sum of Squared Deviations	Degree of Freedom	Mean Square	F-Value	$F_{0.05}$	p-Value	Significance
A	0.0321	4	0.0080	1.1911	3.260	0.3639	Not significant
B	0.1507	4	0.0377	5.5879	3.260	0.0089	Significant
C	0.1274	4	0.0319	4.7245	3.260	0.0160	Significant
Error	0.0028	12	0.0002				

In the above analysis of variance table, the significance level is $\alpha = 0.05$ (confidence level of 95%), and the calculated F-critical value at $F_{0.05}$ is 3.260.

According to Table 8, if the F-means of the pore permeability variation amplitude (A) and temperature (C) factors are greater than $F_{0.05}$, then the pore permeability variation amplitude (A) and temperature (C) have a significant impact on the thermal diffusion coefficient of reservoir rocks at a confidence level of 95%. According to Table 9, the F-values of the pore permeability variation amplitude (A), water saturation (B), and temperature (C) factors are all greater than $F_{0.05}$. Therefore, at a confidence level of 95%, the pore permeability variation amplitude (A), water saturation (B), and temperature (C) have a significant impact on the thermal conductivity of reservoir rocks. According to Table 10, if the F-values of water saturation (B) and temperature (C) factors are greater than $F_{0.05}$, this indicates that water saturation (B) and temperature (C) have a significant impact on the specific heat of reservoir rocks at a confidence level of 95%.

Based on the experimental results and literature research results, we analyzed the mechanism of changes in reservoir rock thermal properties under the influence of multiple factors. The amplitude of pore permeability changes has a significant impact on the thermal diffusion coefficient and thermal conductivity of reservoir rocks. This is mainly because when there are more pores in the reservoir rocks, the fluid in the pores will occupy a larger proportion of space and form a continuous phase. At this point, more resistance and dispersion effects need to be overcome when transferring energy inside the medium, thereby slowing down the speed of heat transfer. Therefore, as the porosity and permeability increase, the thermal diffusion coefficient and thermal conductivity of reservoir rocks will decrease. Water saturation has a significant impact on the thermal conductivity and specific heat of reservoir rocks, mainly because water has a higher specific heat and thermal conductivity. As the water saturation of reservoir rocks increases, their average thermal conductivity and specific heat will gradually increase. It should be noted that in practical situations, the influence of water saturation on the thermal conductivity and specific heat of reservoir rocks of different types, pore structures, and permeability may also vary. For example, when the pore connectivity of reservoir rocks is good and the morphology is regular, the fluid flow in them is better, and the thermal conductivity of reservoir rocks is also better. When the pores of reservoir rocks are small, dispersed, or fractured, this will affect the contact area between liquid water and solid rocks, thereby reducing the total specific heat of reservoir rocks. Temperature has a significant impact on the thermal diffusion coefficient, thermal conductivity, and specific heat of reservoir rocks, mainly because as the temperature increases, the internal microstructure and oil-water saturation of the rocks change. Firstly, as the temperature increases, the vibration frequency of molecules and atoms inside the rock increases, which enhances the rate of energy transfer and diffusion. In theory, the thermal conductivity and thermal diffusion coefficient should increase with the increase in temperature. However, in reality, the small pores and cracks inside the rock can form thermal barriers, hindering the transfer and diffusion of heat. Therefore, the thermal conductivity and thermal diffusion coefficient will actually decrease with the increase in temperature. In addition, as the temperature increases, atoms and molecules in the rock begin to vibrate more violently, leading to an increase in the interaction force between atoms and molecules in the reservoir rock. This interaction force creates a tendency in the rock to resist external changes; that is, the specific heat of the reservoir rock increases.

3.2. Regression Analysis

After determining the impact trend and significance of each factor on different indicators through intuitive analysis and an analysis of variance, regression analysis of experimental data is also necessary to establish regression equations to characterize the quantitative relationship between each factor and different indicators. When conducting regression analysis on experimental data, multiple regression is used to characterize the relationship between each variable and the dependent variable, with the dependent variables being Y_1 (thermal diffusion coefficient), Y_2 (thermal conductivity), and Y_3 (specific heat), while the independent variables are X_A (pore permeability change amplitude), X_B (water saturation), and X_C (temperature).

3.2.1. Multiple Linear Regression

A multiple linear regression method was used to perform regression analysis on the data, and the regression results analysis table is as follows.

According to Table 11, the fitting degree of the regression equations for the thermal diffusion coefficient and thermal conductivity coefficient obtained by the multiple linear regression method is relatively low (below 0.9). Although their *p*-values meet the significance test requirements, the fitting degree of the regression equation is poor and cannot meet the accuracy requirements of the prediction results. The regression equation of specific heat obtained from multiple linear regressions has a high degree of fit (greater than 0.95), and the *p*-value meets the significance test requirements. This indicates that the linear regression equation has a good fit and can meet the accuracy requirements of the prediction results.

Table 11. Multivariate linear regression analysis table.

	Regression Equation	**R^2**	**F-Value**	***p*-Value**
Thermal diffusion coefficient	$Y_1 = 0.7554 - 0.00604X_A + 0.087X_B - 0.00091X_C$	4	0.0080	1.1911
Thermal conductivity	$Y_2 = 1.4388 - 0.01884X_A + 0.268X_B - 0.001X_C$	4	0.0377	5.5879
Specific heat	$Y_3 = 1.8318 - 0.00496X_A + 0.273X_B + 0.00084X_C$	4	0.0319	4.7245

The results of multiple linear regressions indicate that there is no non-linear relationship between the three factors involved in this article and the thermal diffusion coefficient and thermal conductivity coefficient, and that linear regression cannot be used for quantitative analysis. The regression equation of specific heat obtained from multiple linear regression meets the requirements of fitting and significance, and a linear relationship can be considered to characterize the quantitative relationship between specific heat and the three factors involved in this article.

3.2.2. Multiple Nonlinear Regression

The multiple nonlinear regression method used in References [22,23] was used to perform multiple nonlinear regression on the experimental results. The Levenberg–Marquardt method was chosen as the estimation method, and the regression results are shown below (Table 12).

According to Table 13, the fitting degree of the multiple nonlinear regression equations for the three indicators is relatively high (greater than 0.99), and their *p*-values are far less than 0.05. This indicates that the results of multiple nonlinear regressions are superior to those of multiple linear regression in terms of fitting degree and significance.

Table 12. Multivariate nonlinear regression equation coefficient table.

Coefficient	Thermal Diffusion Coefficient Equation	Thermal Conductivity Equation	Specific Heat Equation
n	1	2	3
a	0.966	1.776	1.832
b	−10,293.401	10,301.728	−10,271.272
c	10,293.402	−10,301.727	10,271.273
d	4.234×10^{-6}	5.533×10^{-6}	1.164×10^{-6}
e	0.031	0.035	0.011
f	2.762×10^{-5}	-8.954×10^{-5}	-3.524×10^{-5}
g	0.001	1.11×10^{-4}	−0.001
h	−0.032	−0.045	−0.022
i	−1.41	−0.344	−0.221
j	−0.001	−0.01	−0.002

Table 13. Multivariate nonlinear regression analysis table.

	Regression Equation	R^2	F-Value	*p*-Value
Thermal diffusion coefficient	$Y_n = a + bX_A^2 + cX_B^2 + dX_C^2 + eX_AX_B + fX_AX_C + gX_BX_C + hX_A + iX_B + jX_C$	0.999	7.89×10^4	1.59×10^{-33}
Thermal conductivity		0.991	6211.5	3.02×10^{-25}
Specific heat		0.998	1.78×10^5	3.6×10^{-13}

From the analysis results, it can be seen that the influence direction and significance of the three factors of pore permeability change amplitude, water saturation, and temperature on the thermal diffusion coefficient, thermal conductivity coefficient, and specific heat of reservoir rocks are different. Moreover, during the rapid preheating process of SAGD, reservoir expansion and the steam circulation process will cause simultaneous changes in pore permeability, oil-water saturation, and the temperature of reservoir rocks. Therefore, it is only by analyzing the changes in thermal properties of reservoir rocks and establishing corresponding multiple nonlinear regression equations to characterize them before conducting SAGD rapid preheating operations that we can more accurately predict the changes in reservoir rocks during production and construction and can further evaluate the effectiveness of SAGD rapid preheating construction more accurately.

3.3. Prediction Model for the Thermal Properties of Reservoir Rocks

The main controlling factors, significance, and regression equations affecting the thermal diffusion coefficient, thermal conductivity coefficient, and specific heat of reservoir rocks have been determined through intuitive analysis, variance analysis, and regression analysis in the previous section. Taking into account the accuracy and significance of the regression results, a multiple nonlinear regression equation was ultimately chosen to characterize the quantitative relationship between the thermal diffusion coefficient, thermal conductivity coefficient, and specific heat indicators and the changes in porosity and permeability, water saturation, and temperature.

The regression equation for thermal diffusion coefficient is as follows:

$$\begin{aligned} \alpha = {} & 0.966 - 10,293.401x^2 + 10,293.402S_W^2 + 4.234 \times 10^{-6}T_f^2 + 0.031xS_W \\ & + 2.762 \times 10^{-5}xT_f + 0.001S_WT_f - 0.032x + -1.41S_W - 0.001T_f \end{aligned} \tag{1}$$

The regression equation for thermal conductivity is as follows:

$$\begin{aligned} \lambda_f = {} & 1.776 + 10,301.728x^2 - 10,301.727S_W^2 + 5.533 \times 10^{-6}T_f^2 + 0.035xS_W \\ & - 8.954 \times 10^{-5}xT_f + 1.11 \times 10^{-4}S_WT_f - 0.045x + (-0.344S_W) - 0.01T_f \end{aligned} \tag{2}$$

The regression equation for specific heat is as follows:

$$M_R = 1.832 - 10,271.272x^2 + 10,271.273S_W^2 + 1.164 \times 10^{-6}T_f^2 + 0.011xS_W \\ -3.524 \times 10^{-5}xT_f - 0.001S_WT_f - 0.022x + -0.221S_W - 0.002T_f \tag{3}$$

α—Thermal diffusion coefficient of reservoir rocks, mm^2/s
x—Increase in pore permeability, %
S_W—Reservoir rock water saturation, dimensionless
T_f—Reservoir rock temperature, °C
λ_F—Thermal conductivity coefficient of reservoir rock, W/(m·K)
M_R—specific heat of reservoir rocks, J/(g·K)

4. Conclusions

(1) The main factor affecting the thermal diffusion coefficient of reservoir rocks is the amplitude of pore permeability changes. The main factor affecting the thermal conductivity of reservoir rocks is temperature. The main factor affecting the specific heat of reservoir rocks is water saturation.

(2) The thermal properties of reservoir rocks are influenced by the amplitude of pore permeability changes, water saturation, and temperature. An increase in porosity and permeability will cause the fluid to form a continuous phase, resulting in the need to overcome more resistance and dispersion when transferring energy within the medium, thereby slowing down the rate of heat transfer. The specific heat and thermal conductivity of water are relatively high. As the water saturation of reservoir rocks increases, the average thermal conductivity and specific heat of reservoir rocks will gradually increase. An increase in temperature will increase the vibration frequency of molecules and atoms inside the rock, increasing the rate of energy transfer and diffusion. However, the small pores and cracks present in the rock will form thermal barriers, which, in turn, reduce the thermal conductivity and thermal diffusion coefficient. In addition, the specific heat of reservoir rocks increases due to the increased temperature, which enhances the interaction forces between atoms and molecules in the reservoir rocks.

(3) Taking into account the accuracy and significance of the regression results, a multiple nonlinear regression equation was ultimately chosen to characterize the quantitative relationship between reservoir thermal properties and pore permeability changes, water saturation, and temperature.

Author Contributions: J.T., W.Y. and Z.Q. conceived and designed the experiments; S.H., M.D. and J.T. conducted the experiments and analyzed the data; J.T. wrote the paper. All authors have read and agreed to the published version of the manuscript.

Funding: This research was funded by the National Natural Science Foundation of China (grant number 52004048, U22B2074), the Natural Science Foundation of Chongqing Municipality, China (grant number cstc2020jcyj msxmX0856), and the Chongqing Science and Technology Bureau Chongqing Talent Plan Contract System Project (cstc2022ycjh-bgzxm0055).

Data Availability Statement: The data that support the findings of this study are available from the corresponding author.

Conflicts of Interest: The authors declare no conflicts of interest.

References

1. Xu, X. On the Factors and Mechanisms Influencing the Thermo physical Properties of Rocks. *Pet. Explor. Dev.* **1992**, *6*, 85–89.
2. Cheng, C.; Yu, W.; Jia, W.; Lin, H.; Li, L. Research progress and development trends of rock thermo physical properties. *Prog. Earth Sci.* **2017**, *10*, 1072–1083.
3. Sun, Q.; Lü, C.; Cao, L.; Li, W.; Geng, J.; Zhang, W. Thermal properties of sandstone after treatment at high temperature. *Int. J. Rock Mech. Min. Sci.* **2016**, *85*, 60–66. [CrossRef]
4. Abdulagaov, I.M.; Abdulagatova, Z.Z.; Kallaev, S.N.; Omarov, Z.M. Heat-capacity measurements of sandstone at high temperatures. *Geomech. Geophys. Geo-Energy Geo-Resour.* **2019**, *1*, 65–85. [CrossRef]

5. Emirov, S.N.; Aliverdiev, A.A.; Zarichnyak, Y.P.; Emirov, R.M. Studies of the Effective Thermal Conductivity of Sandstone Under High Pressure and Temperature. *Rock Mech. Rock Eng.* **2021**, *54*, 3165–3174. [CrossRef]
6. Geng, J.; Sun, Q.; Zhang, Y.; Cao, L.; Lü, C.; Zhang, Y. Temperature dependence of the thermal diffusivity of sandstone. *J. Pet. Sci. Eng.* **2018**, *164*, 110–116. [CrossRef]
7. Liu, Z.; Wang, L.; Zhao, B.; Leng, J.; Zhang, G.; Yang, D. Heat Transfer in Sandstones at Low Temperature. *Rock Mech. Rock Eng.* **2019**, *1*, 35–45. [CrossRef]
8. Guo, P.; Bai, B.; Chen, S.; Shi, C.; Du, H. Experimental study on the influence of temperature and water content on the thermal conductivity characteristics of sandstone. *J. Rock Mech. Eng.* **2017**, *S2*, 3910–3914.
9. Hu, R.; Wang, Y.; Xiao, X.; Chen, M.; Chen, Q. Research and Application of Rock Thermophysical Parameters in Chunguang Oilfield. *Pet. Geol. Eng.* **2017**, *6*, 108–111.
10. Song, X.; Jiang, M.; Peng, Q.; Xiong, P. Analysis of thermal properties parameters and influencing factors of major rock formations in Guizhou. *J. Geol.* **2019**, *8*, 2092–2103.
11. Zhen, Z.; Zhu, J.; Yu, G.; Gai, Y.; Zhang, H. Experimental Study on Thermal Physical Parameters of Surrounding Rock in Lanzhou Underground Rail Transit. *Mod. Tunn. Technol.* **2020**, *2*, 163–168.
12. Schärli, U.; Rybach, L. On the thermal conductivity of low-porosity crystalline rocks. *Tectonophysics* **1984**, *103*, 307–313. [CrossRef]
13. El Sayed, A.A.A. Thermophysical study of sandstone reservoir rocks. *J. Pet. Sci. Eng.* **2011**, *3*, 138–147. [CrossRef]
14. Zhu, W.; Su, X.; Liu, Q. Analysis of the relationships between the thermophysical properties of rocks in the Dandong Area of China. *Eur. J. Remote Sens.* **2020**, *54*, 122–131. [CrossRef]
15. Huang, J.H. Effective thermal conductivity of porous rocks. *J. Geophys. Res.* **1971**, *76*, 6420–6427. [CrossRef]
16. Esser, A.; Redmer, R.; Röpke, G. Interpolation formula for the electrical conductivity of nonideal plasmas. *Contrib. Plasma Phys.* **2003**, *43*, 33–38. [CrossRef]
17. Torquato, S. Thermal Conductivity of Disordered Heterogeneous Media from the Microstructure. *Rev. Chem. Eng.* **1987**, *4*, 151–204. [CrossRef]
18. Litovskii, E.Y. Interpolation formula for the expression of thermal conductivity of solid materials on porosity. *Izv. Akad. Nauk. Sssr Seriianeorganicheskie Mater.* **1980**, *16*, 559–560.
19. Buntebarth, G.; Schopper, J. Experimental and theoretical investigations on the influence of fluids, solids and interactions between them on thermal properties of porous rocks. *Phys. Chem. Earth* **1998**, *23*, 1141–1146. [CrossRef]
20. Aichlmayr, H.T. The Effective Thermal Conductivity of Saturated Porous Media. Master's Thesis, University of Minnesota, Minneapolis, MN, USA, 1999.
21. Luo, J.; Jia, J.; Zhao, H.; Zhu, Y.; Guo, Q.; Cheng, C.; Tan, L.; Xiang, W.; Rohn, J.; Blum, P. Determination of the thermal conductivity of sandstones from laboratory to field scale. *Environ. Earth Sci.* **2016**, *75*, 1158. [CrossRef]
22. Zhang, Q.; Xie, Y.; Wu, A. Prediction of airborne conformal antenna absorbing coating thickness based on orthogonal experiments and regression analysis. *J. Ordnance Ind.* **2021**, *12*, 2693–2699.
23. Lü, S.; Ni, W.; Army, N.R.; Wang, X. Research on Aluminum Ash Hydrolysis Based on Orthogonal Experiments and Nonlinear Regression Analysis. *Nonferrous Met. Eng.* **2019**, *10*, 52–56.

Disclaimer/Publisher's Note: The statements, opinions and data contained in all publications are solely those of the individual author(s) and contributor(s) and not of MDPI and/or the editor(s). MDPI and/or the editor(s) disclaim responsibility for any injury to people or property resulting from any ideas, methods, instructions or products referred to in the content.

energies
MDPI

Article

Comparison of the Reaction Characteristics of Different Fuels in the Supercritical Multicomponent Thermal Fluid Generation Process

Qiang Fu [1], Jie Tian [2,*], Yongfei Liu [1], Zhilin Qi [2,*], Hongmei Jiao [1] and Shenyao Yang [2]

1 State Key Laboratory of Natural Gas Hydrates, Beijing 100028, China; fuqiang8@cnooc.com.cn (Q.F.)
2 School of Petroleum and Natural Gas Engineering, Chongqing University of Science & Technology, Chongqing 401331, China; yangshenyaocq@163.com
* Correspondence: tianjie@cqust.edu.cn (J.T.); 2008008@cqust.edu.cn (Z.Q.)

Abstract: Supercritical multicomponent thermal fluid technology is a new technology with obvious advantages in offshore heavy oil recovery. However, there is currently insufficient understanding of the generation characteristics of the supercritical multicomponent thermal fluid, which is not conducive to the promotion and application of this technology. In order to improve the economic benefits and applicability of the supercritical multicomponent thermal fluid thermal recovery technology, this article reports on indoor supercritical multicomponent thermal fluid generation experiments and compares the reaction characteristics of different fuels in the supercritical multicomponent thermal fluid generation process. The research results indicate that the main components of the products obtained from the supercritical water–crude oil/diesel reaction are similar. Compared to the supercritical water–crude oil reaction, the total enthalpy value of the supercritical multicomponent thermal fluid generated by the supercritical water–diesel reaction is higher, and the specific enthalpy is lower. When the thermal efficiency of the boiler is the same, the energy equilibrium concentration of crude oil is lower than that of diesel. The feasibility of using crude oil instead of diesel to prepare supercritical multicomponent thermal fluids is analyzed from three aspects: reaction mechanism, economic benefits, and technical conditions. It is believed that using crude oil instead of diesel to prepare supercritical multicomponent thermal fluids has good feasibility.

Keywords: heavy oil; supercritical multicomponent thermal fluid; thermal recovery

1. Introduction

China has abundant heavy oil resources, with offshore heavy oil resources mainly concentrated in the Bohai Bay. The conventional heavy oil thermal recovery technology has achieved good results on land, but its adaptability is poor in sea [1–5]. The problems are mainly concentrated in the following aspects: First, conventional thermal recovery technologies mostly use steam as the heat carrier, which has high requirements for water quality. The water treatment process on offshore platforms is complex, and the costs for water treatment and pipeline maintenance are very high. Second, the preparation of steam heavily relies on diesel fuel, which is costly and emits a large amount of carbon dioxide. Third, there is severe heat loss during the steam injection process, resulting in low thermal efficiency.

Therefore, in 2017, Zhou et al. proposed the supercritical multicomponent thermal fluid technology [6]. Supercritical multicomponent thermal fluids are a type of mixed fluid generated by the reaction of supercritical water with diesel or crude oil, primarily consisting of supercritical water, nitrogen, and carbon dioxide. Due to the strong reactivity of supercritical water, supercritical multicomponent thermal fluid exhibits the following three characteristics: Firstly, it has high solubility for organic compounds and gases in oil and can approximate miscible flooding, improving the efficiency of heavy oil displacement;

Secondly, it has high diffusivity, low surface tension and dynamic viscosity, large swept volume after injection into the formation, and a good displacement effect; Thirdly, it has high reactivity and can significantly reduce the conversion temperature of long-chain macromolecular hydrocarbons in heavy oil, thereby improving the properties of heavy oil in the formation and enhancing its flow capacity.

Supercritical multicomponent thermal fluids are primarily composed of supercritical water, carbon dioxide (CO_2), and nitrogen (N_2). Their production predominantly utilizes techniques such as supercritical water gasification and supercritical water oxidation. Leveraging the exceptional solubility and diffusivity of supercritical water, a wide array of organic waste liquids, including diesel and oily effluents, can be completely dissolved and rapidly vaporized under the extreme temperature and pressure characteristics of supercritical conditions. This process converts them into gasification products primarily consisting of hydrogen (H_2) and carbon dioxide (CO_2). Subsequently, these products are combusted with oxygen-enriched gases that are also dissolved in supercritical water, culminating in the formation of supercritical multisource, multicomponent thermal fluids. This approach harnesses the unique properties of supercritical water to efficiently transform waste into a potent source of energy [6–10]. During the supercritical water gasification process, this unique medium significantly enhances the conversion efficiency of the feedstock. It diminishes the coke formation within the system and exerts a profound impact through its solvent and dispersive capabilities, hydrogen donation, and acid-catalytic functions. These attributes lead to remarkable desulfurization, denitrification, and the removal of heavy metals, thereby contributing to a cleaner and more efficient gasification outcome. The transformative power of supercritical water lies in its ability to act as a potent solvent that breaks down complex organic molecules, while its capacity to transfer hydrogen and catalyze reactions accelerates the gasification process, yielding a more refined end product [11,12].

In the oxidation process of supercritical water, higher content of organic matter brings greater heat release [13–16]. Additionally, once the concentration of organic material in the reaction mixture crosses a specific threshold, the exothermic nature of the reaction can be harnessed to maintain the heat balance, thereby enabling an energy-self-sustaining process. This innovative approach significantly enhances the economic viability of the entire operation. Preliminary findings from our research indicate that an oxygen surplus exceeding 10% is sufficient to ensure the complete dissociation of all substances involved in the reaction. Beyond this point, an increase in oxygen levels does not exert any additional influence on the reaction's progress [17]. Furthermore, it is observed that when the organic matter concentration in the supercritical water oxidation reaction exceeds 2%, the reaction becomes self-heating. This is attributed to the substantial heat liberated during the reaction process, which is sufficient to sustain the reaction's temperature requirements without the need for external heating sources. This self-sustaining characteristic is a significant advantage, as it contributes to the overall energy efficiency and cost-effectiveness of the process [18–21]. Experiments on wastewater with different chemical oxygen consumption ranges show that as long as the chemical oxygen consumption ranges from 183 to 437 g/L and the mass flow rate ranges from 20.83 to 104.17 kg/h, the reaction can be maintained entirely by the heat released from its own reaction [21].

Previous work mainly focuses on the supercritical water treatment of wastewater, and the research on the reaction mechanism of supercritical water and the products after the reaction is relatively mature. However, there is currently little discussion on using different fuels to generate supercritical multicomponent thermal fluids under different conditions, let alone comparing the similarities and differences of different fuels in the process of generating supercritical multicomponent thermal fluids and the substitutability between fuels. However, these are precisely the things that oil field engineers are very concerned about. In view of this and to better understand the process of generating supercritical multicomponent thermal fluids using different fuels and clarify the differences brought about by various fuels, in this study, we conducted experiments on the generation of supercritical multicomponent thermal fluids using diesel and crude oil under different conditions. Based on the experimental results, the characteristics of supercritical multicomponent thermal fluids prepared using different fuels under different conditions were compared and analyzed, and the feasibility of using crude oil instead of diesel to generate supercritical multicomponent thermal fluids was tested. This study marks the first comprehensive comparison of the differences in generating supercritical multicomponent thermal fluids using various fuels. It holds significant guiding importance for the rational selection of fuels for the generation of supercritical multicomponent thermal fluids.

2. Materials and Methods

(1) Experimental materials

The diesel used in these experiments is 0# diesel, and the crude oil is from the L block of Bohai Bay, China. At reservoir temperature (50 °C), the viscosity of crude oil is 1756 MPa·s, which belongs to ordinary heavy oil. The SARA analysis results of crude oil are shown in Table 1.

Table 1. SARA analysis results of crude oil.

No.	Component	Wt.%
1	Asphaltenes	25.99
2	Resins	8.14
3	Aromatics	19.65
4	Saturates	46.22
5	Total	100.00

(2) Apparatus

As illustrated in Figure 1, the experimental setup is anchored by a supercritical multicomponent thermal fluid generation system and a chromatographic analysis apparatus. The former includes a high-temperature, high-pressure reactor (as shown in Figure 2), an ISCO pump, a check valve, and an intermediate holding vessel, all meticulously designed to facilitate the generation process. The chromatographic analysis apparatus, on the other hand, is equipped with both a gas chromatograph and a liquid chromatograph, ensuring a comprehensive analysis of the thermal fluid components. This sophisticated arrangement of equipment ensures a seamless workflow from the production of supercritical fluids to their detailed characterization.

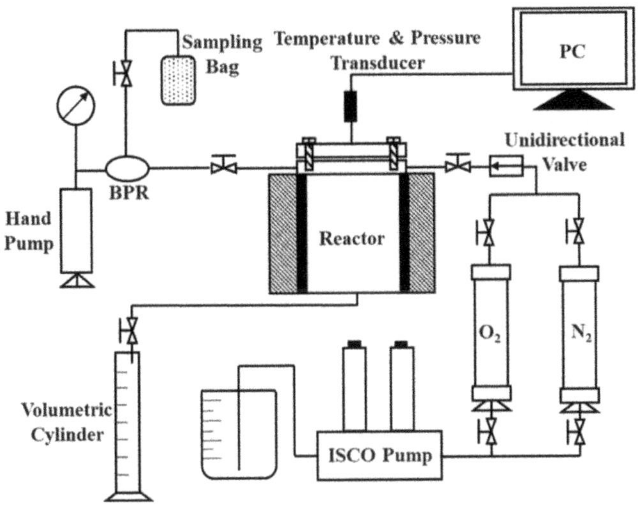

Figure 1. Experimental flow of supercritical multicomponent thermal fluid generation.

Figure 2. High temperature and pressure reactor.

(3) Procedures

①Analyzing diesel oil/crude oil samples by liquid chromatography before the experiment.

② Checking the status of experimental equipment and repairing faulty components.

③ Connecting the experimental equipment according to the experimental flow chart.

④ Carefully measuring diesel oil or crude oil sample and water according to the predetermined ratio; the mixture is introduced into the high-temperature, high-pressure reactor, ensuring a secure seal to contain the process within. The experimental vessel is crafted from HC276 alloy, a material renowned for its durability and resistance to extreme conditions. It is designed to withstand a maximum working pressure of up to 40 MPa and can operate effectively at temperatures reaching as high as 600 °C, providing a robust platform for conducting experiments under demanding thermal and pressure regimes.

⑤ Opening the air inlet and outlet, charging nitrogen for more than 10 min through the air inlet with a large displacement to ensure that the air in the reactor is cleared, and then closing the air outlet.

⑥ Filling nitrogen into the reactor up to the design pressure and then turning on the reactor power supply and the heating mode.

⑦ Upon attaining the supercritical state within the reactor, as indicated by the optimal temperature and pressure thresholds, the heating process is gradually discontinued, and the agitation function is engaged. Throughout the entirety of the reaction experiment, the reactor is continuously supplied with oxygen, ensuring a steady progression of the reaction to its conclusion. Concurrently, the fluctuating temperatures and pressures within the reactor during the experiment are meticulously monitored and documented in real time, capturing the dynamic essence of the supercritical transformation.

⑧ Once the reaction is complete, the reactor is allowed to cool, descending from its supercritical temperatures to a safe, ambient level. As the reactor's temperature returns to normalcy, the air outlet is cautiously opened, facilitating the collection of gas samples that have been formed within the vessel. Subsequently, the pressure within the reactor is gradually relieved, and the reactor itself is opened to carefully harvest any remaining liquid and solid residues, if present, ensuring that all components are accounted for in the post-reaction analysis.

⑨ Analyzing the collected gas–liquid samples by chromatography.

The parameters determined for the experiment are shown in Table 2. Given the specific conditions on the oil field, two types of fuels, three temperature conditions, and two pressure conditions are being considered.

Table 2. Experimental parameters of supercritical water–oil reaction.

Case	Reactant	Initial Temperature and Pressure
1	45 mL water + 5 mL diesel oil	400 °C/23 MPa
2	45 mL water + 5 mL diesel oil	400 °C/25 MPa
3	45 mL water + 5 mL diesel oil	450 °C/25 MPa
4	45 mL water + 5 mL diesel oil	500 °C/25 MPa
5	45 mL water + 5 mL diesel oil	500 °C/23 MPa
6	45 mL water + 5 mL diesel oil	400 °C/25 MPa
7	28.3 mL water + 5 mL diesel oil	400 °C/25 MPa
8	20 mL water + 5 mL diesel oil	400 °C/25 MPa
9	15 mL water + 5 mL diesel oil	400 °C/25 MPa
10	11.67 mL water + 5 mL diesel oil	400 °C/25 MPa
11	45 mL water + 5 mL crude oil	400 °C/23 MPa
12	45 mL water + 5 mL crude oil	400 °C/25 MPa
13	45 mL water + 5 mL crude oil	450 °C/25 MPa
14	45 mL water + 5 mL crude oil	500 °C/25 MPa
15	45 mL water + 5 mL crude oil	500 °C/23 MPa
16	45 mL water + 5 mL crude oil	400 °C/25 MPa
17	28.3 mL water + 5 mL crude oil	400 °C/25 MPa
18	20 mL water + 5 mL crude oil	400 °C/25 MPa
19	15 mL water + 5 mL crude oil	400 °C/25 MPa
20	11.67 mL water + 5 mL crude oil	400 °C/25 MPa

3. Results and Discussion

3.1. Comparison of Reaction Product Composition, Product State, and Reaction Mechanism of Supercritical Water–Crude Oil and Water–Diesel Oil

Following the completion of the supercritical water–oil (encompassing diesel oil and crude oil) reaction experiment, which was conducted under various oil–water ratios and a range of temperature and pressure settings, the reactor was allowed to cool down to ambient temperature. Subsequently, the gaseous contents within the reactor were extracted and subjected to chromatographic analysis. This process was undertaken to ascertain the precise makeup of the supercritical multicomponent thermal fluid.

The chromatographic analysis results (Figures 3 and 4) indicate that the primary constituents of the gaseous byproducts from the reactions involving diesel oil and heavy oil (at room temperature) are largely identical. This implies that the primary components are carbon dioxide, nitrogen, and oxygen that were not utilized in the reaction process. However, when contrasting the reaction outputs of diesel oil with those of heavy oil, it is evident that the latter reaction yields a higher concentration of impurity gases, including sulfur dioxide and nitrogen dioxide. This disparity is attributed to the fact that heavy oil, unlike diesel oil, contains asphaltene, which is composed of heteroatom elements, such as sulfur and nitrogen. These elements, upon undergoing complete combustion in an oxygen-rich environment, transform into sulfur dioxide and nitrogen dioxide.

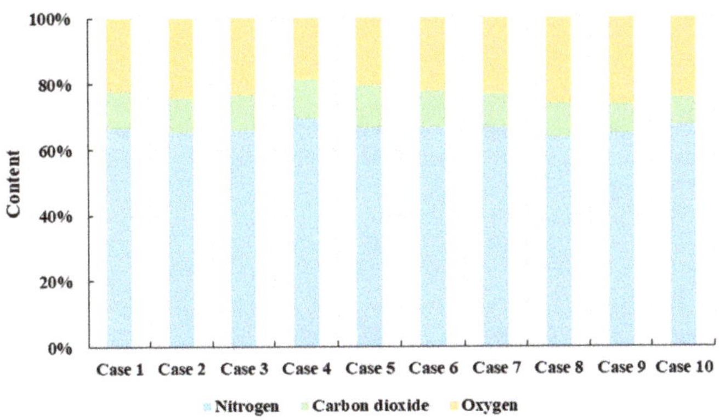

Figure 3. Gas chromatography analysis of supercritical water–diesel oil reaction products.

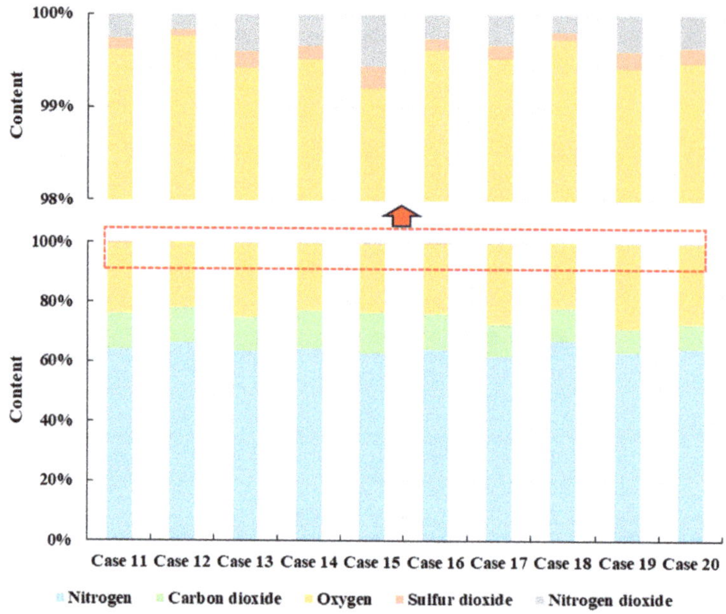

Figure 4. Gas chromatography analysis of supercritical water–crude oil reaction products.

The influence of the oil–water ratio in the reactants is more pronounced on the chromatographic outcomes. In the experiments, while the overall quantity of oil was kept constant, the oil–water ratio was modified by varying the volume of water. Given a fixed reactor size, the amount of nitrogen required to achieve the same initial pressure varied depending on the oil–water ratio. Specifically, a higher oil proportion, which corresponds to a lower water volume, results in a smaller combined volume of oil and water. Consequently, more nitrogen is needed to reach the same starting pressure compared to other experiments with different oil–water ratios. As a result, the examination of the chromatograms from the gases collected post-reaction reveals a clear trend: as the oil ratio increases, the concentration of carbon dioxide diminishes, and the concentration of nitrogen increases. Since the quantity of oil in the reaction is constant, the amount of oxygen introduced is roughly equivalent across experiments. Thus, the residual oxygen's proportion in the chromatographic results remains essentially consistent.

The reaction between supercritical water and diesel oil as well as the reaction between supercritical water and crude oil can be roughly divided into two stages, i.e., the supercritical water first disperses the continuous oil, and then the oil is oxidized and reacted completely under the joint action of supercritical water and oxygen, and intermediates will be generated in the oxidation process. However, since crude oil has more long-chain compounds, it is not only physically dispersed in the first stage of reaction with supercritical water but also has an obvious cracking process, which is the difference between the supercritical water–diesel oil reaction and the supercritical water–crude oil reaction mechanisms.

3.2. Enthalpy Comparison of Supercritical Water–Crude Oil/Diesel Oil Reactions to Generate Supercritical Multicomponent Thermal Fluids

Generally speaking, a thermal fluid with greater enthalpy indicates a higher heat transfer capacity, which is more beneficial for the extraction of viscous oil. Thus the enthalpy of the thermal fluid produced through the reaction is a very intriguing metric. It should be noted that due to constraints of the experimental methodology and procedures employed in this study, the quantities of nitrogen and oxygen introduced into the reactor do not mirror the natural ratio found in air. In mining operations, the gas typically introduced into the reaction is air itself. To ensure that the calculated enthalpy of the supercritical multicomponent thermal fluid has broad applicability and is comparable, the enthalpy of the supercritical multicomponent thermal fluid generated in various reactions can be determined using the output fluid chromatography results. This process involves adjusting the nitrogen content to align with the air's nitrogen–oxygen ratio and disregarding the reaction's residual oxygen. The calculation formula is as follows:

$$h = \frac{w_i}{\sum_{i=1}^{n} w_i} h_i \tag{1}$$

$$\overline{h} = \frac{h}{\sum_{i=1}^{n} w_i} \tag{2}$$

where h is the enthalpy of the supercritical multicomponent thermal fluid, J. h_i is the enthalpy of the component i in the supercritical multicomponent thermal fluid, J. w_i is the mass of the component i in the supercritical multicomponent thermal fluids, g. i is any component in the supercritical multicomponent thermal fluids; $\overline{h}$ is the specific enthalpy, which is the enthalpy value of a supercritical multicomponent thermal fluid per unit mass.

Viewing through the lens of specific enthalpy (Figures 5 and 6), an inverse relationship is observed between the proportion of diesel and the specific enthalpy of the thermal fluid—it increases as the diesel content diminishes and the water content grows. This phenomenon can be attributed to the predominant heat-carrying role of supercritical water within the thermal fluid. An elevated proportion of supercritical water correlates with a higher enthalpy in the resultant supercritical multicomponent thermal fluid. Employing the method previously described, we calculated the specific enthalpy values of supercritical multicomponent thermal fluids produced under various initial temperatures and pressures. As alluded to earlier, the initial conditions of temperature and pressure exert a negligible influence on the reaction products' composition. Consequently, when normalized to a uniform temperature and pressure (400 °C, 25 MPa), the total and specific enthalpy values for the thermal fluids, derived from experiments under differing initial states, are strikingly similar. The computed average values are 180,934.940 J for total enthalpy and 1506.448 J/g for specific enthalpy.

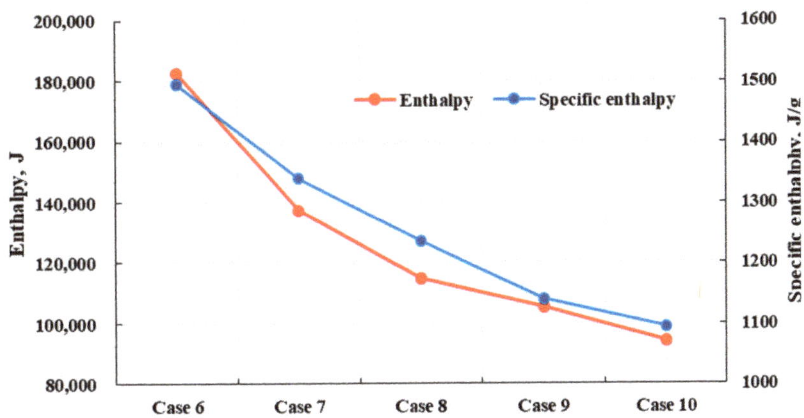

(a)

Figure 5. *Cont.*

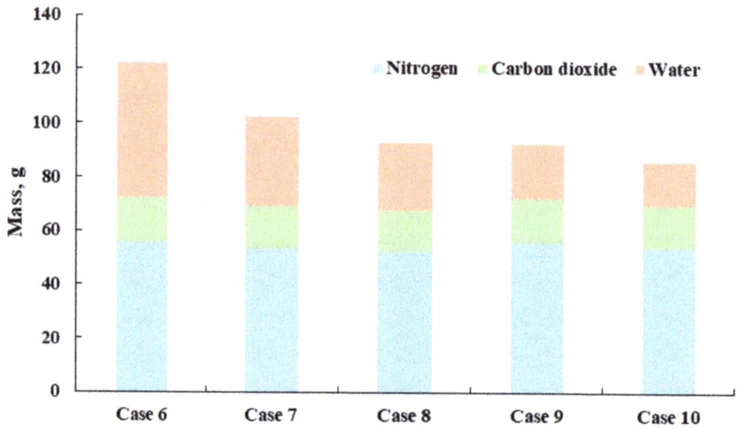

(**b**)

Figure 5. The supercritical multicomponent thermal fluid generated by supercritical water–diesel reaction under different oil–water ratios. (**a**) Enthalpy and specific enthalpy of the supercritical multicomponent thermal fluid; (**b**) Mass composition of supercritical multicomponent thermal fluid.

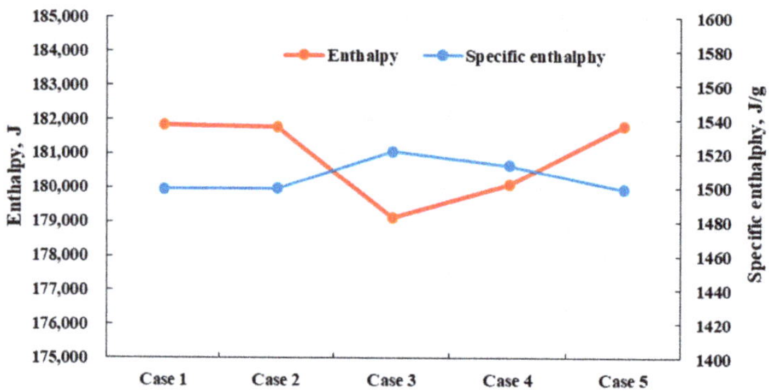

(a)

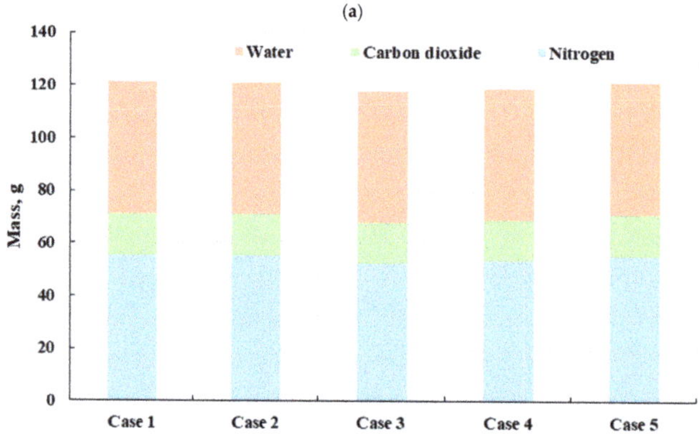

(b)

Figure 6. The supercritical multicomponent thermal fluid generated by supercritical water–diesel reaction under different initial temperature and pressure conditions. (**a**) Enthalpy and specific enthalpy of the supercritical multicomponent thermal fluid; (**b**) Mass composition of supercritical multicomponent thermal fluid.

Consistent with the earlier methodology, the specific enthalpy values of the supercritical multicomponent thermal fluids produced from the reaction between supercritical water and crude oil were calculated based on the composition of the output fluid components. Solely from the perspective of specific enthalpy, the trend is similar to the reaction between supercritical water and crude oil. Specifically, as the water proportion increases, the proportion of supercritical water in the resulting products also rises, leading to higher specific enthalpy values in the thermal fluid.

Using the previously described approach, the specific enthalpy values of supercritical multicomponent thermal fluids produced under varying initial temperature and pressure conditions were determined. When these values are adjusted to a consistent temperature and pressure (400 °C, 25 MPa), the total and specific enthalpy values for the thermal fluids produced under different initial conditions are found to be quite similar, with average totals of 168,630.034 J and 1593.106 J/g, respectively (Figures 7 and 8).

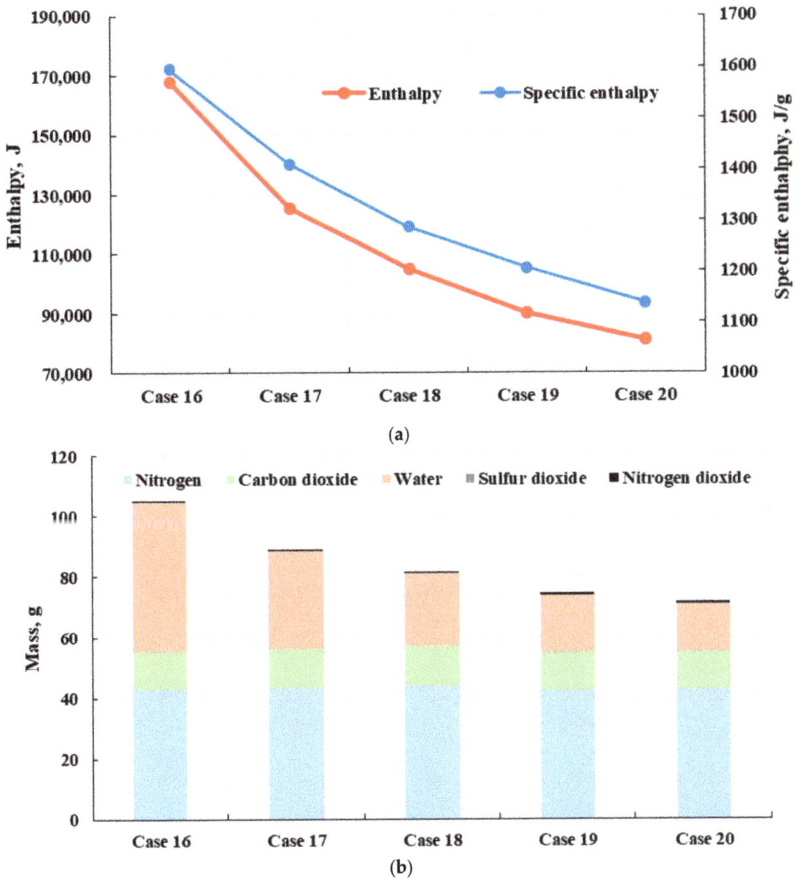

Figure 7. The supercritical multicomponent thermal fluid generated by supercritical water–crude oil reaction under different oil–water ratios. (**a**) Enthalpy and specific enthalpy of the supercritical multicomponent thermal fluid; (**b**) Mass composition of supercritical multicomponent thermal fluid.

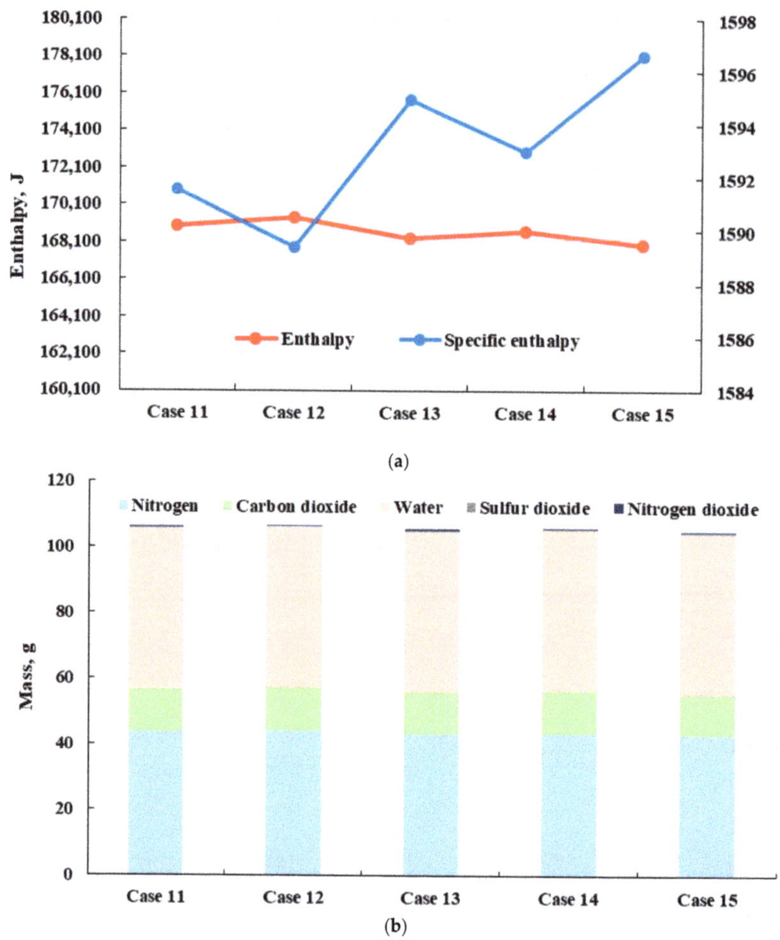

Figure 8. The enthalpy of supercritical multicomponent thermal fluid generated by supercritical water–crude oil reaction under different initial temperature and pressure conditions. (**a**) Enthalpy and specific enthalpy of the supercritical multicomponent thermal fluid; (**b**) Mass composition of supercritical multicomponent thermal fluid.

Analyzing the results from the four preceding figures, it is clear that the total enthalpy of supercritical multicomponent thermal fluids resulting from the reaction between diesel and supercritical water exceeds that of the reaction with heavy oil. Conversely, the specific enthalpy values show an opposite trend. For instance, in the 10% diesel and 10% heavy oil reaction experiments, both initiated at 400 °C and 25 MPa, the diesel reaction yields more carbon dioxide and water, with the excess carbon dioxide being more significant than the excess water. Further analysis indicates that diesel, which is over 95% hydrocarbons with 90% of those being saturated hydrocarbons, has a lower density than heavy oil. The mass of 5 mL of diesel is slightly less than the mass of 5 mL of heavy oil. However, due to the higher oxygen consumption of hydrocarbons, especially the saturated ones, the reaction with diesel consumes more oxygen, leading to an increased production of carbon dioxide and water.

Since oxygen is sourced from air, higher oxygen consumption in the diesel reaction results in a higher nitrogen content in the resulting thermal fluid, which contributes to a greater mass of gas and, consequently, a higher total mass and total enthalpy. On the other hand, heavy oil, with a hydrocarbon content of less than 70% and a saturated hydrocarbon content about half that of diesel, consumes less oxygen during the reaction. This results in lower nitrogen content in the thermal fluid from the heavy oil reaction, leading to less gas mass, total mass, and total enthalpy.

Given that the thermal fluid is predominantly composed of supercritical water and that gases have significantly lower heat capacity than water, their contribution to the overall enthalpy is minor. Therefore, for the supercritical multicomponent thermal fluids produced in the reaction with heavy oil, the proportion of water per unit mass of the thermal fluid is higher, resulting in a higher specific enthalpy compared to those produced in the reaction with diesel (Figures 9 and 10).

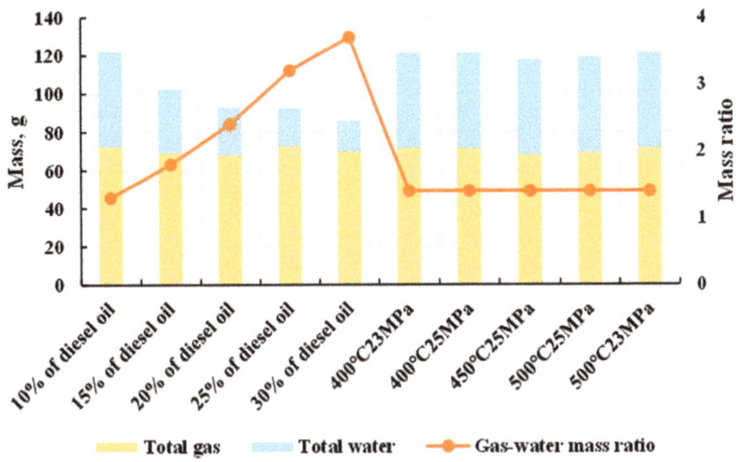

Figure 9. Gas-water mass ratio of supercritical multicomponent thermal fluid generated by supercritical water-diesel reaction under different conditions.

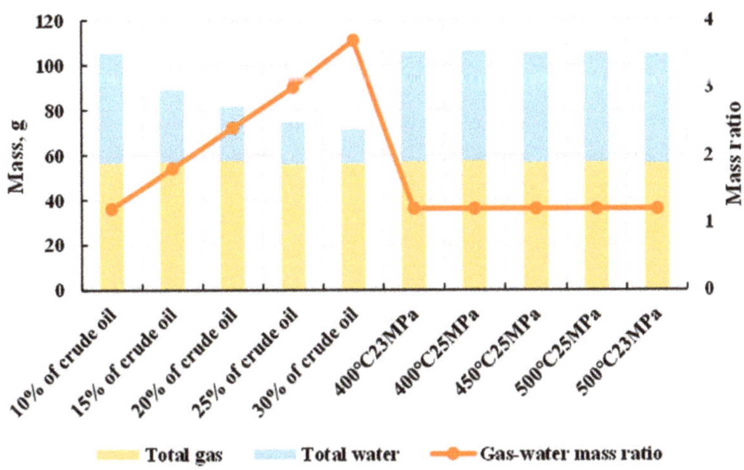

Figure 10. Gas-water mass ratio of supercritical multicomponent thermal fluid generated by supercritical water-crude oil reaction under different conditions.

4. Energy Equilibrium Point in Supercritical Water–Diesel/Crude Oil Reaction Process

In practical applications within oil fields, the preparation of supercritical multicomponent thermal fluids is achieved through the utilization of supercritical multicomponent thermal fluid generators. In this preparation process, the energy balance point of the supercritical water–diesel/crude oil reaction needs to be taken into consideration.

Generally, the energy balance point refers to the equilibrium relationship between the incoming and outgoing heat in the operation of a generator. Typically, the incoming heat in a generator operation is considered to be the low calorific value of the fuel, and the outgoing heat includes the heat utilized in generating steam (or hot water) and the heat lost without utilization.

Based on this definition, for the oil field preparing supercritical multicomponent thermal fluids through a supercritical multicomponent thermal fluid generator, the following statement should uphold: when reaching an energy balance, the heat released per unit time from the supercritical water–diesel/crude oil reaction should be equal to the sum of lost heat (the heat carried by the generated supercritical multicomponent thermal fluid) and the heat absorbed per unit time by the incoming materials heated to the supercritical state inside the generator.

Although several sets of experiments involving supercritical multicomponent thermal fluid–diesel/crude oil reactions have been conducted, no new reactants were added during the laboratory reaction experiments. The heat loss from the laboratory reaction vessel is also irrelevant for the generators used in oil fields. Therefore, the generator thermal efficiency is introduced here to address the issue of heat loss in generators. In other words, when achieving energy balance, the heat released per unit time from the supercritical water–diesel/crude oil reaction multiplied by the generator thermal efficiency should be equal to the sum of the heat carried by the generated supercritical multicomponent thermal fluid and the heat absorbed per unit time by the incoming materials heated to the supercritical state inside the generator. This relationship can be expressed as follows:

$$\mathrm{E} \cdot \eta = H + \Delta h \tag{3}$$

where E is the total heat dissipation of fuel. η is the generator's thermal efficiency. H is the enthalpy of the generated supercritical multicomponent thermal fluid. Δh is the heat absorbed by the new material when heated to the reaction condition.

Since the calculation of this method requires trial calculation, the calculation flow is shown in Figure 11:

Following this approach and computational procedure, calculations were conducted for both the supercritical water-diesel reaction and the supercritical water-crude oil reaction. The thermal efficiency of the generator was varied across five scenarios: 0.5, 0.6, 0.7, 0.8, and 0.9. Utilizing the energy balance relationship mentioned above, the concentrations of diesel and crude oil at the energy balance point can be computed. This provides guidance for the efficient preparation of supercritical multicomponent thermal fluids in oil fields. The parameter values and calculation results are listed below.

From Figure 12, it is evident that despite the similar combustion heats of the crude oil and diesel used, the energy balance concentration for crude oil is lower than that for diesel when the generator efficiency is the same. In other words, if the crude oil from this project is used to prepare supercritical multicomponent thermal fluids in oil fields, the required concentration would be lower than that of diesel, meaning less quantity would be needed.

It is important to note that the energy balance is influenced by various factors. Different fuels will alter the combustion heat, consequently changing the energy balance point. Changes in the generator or generator efficiency will also impact the energy balance point. Moreover, alterations in the temperature and pressure of the output thermal fluid will result in changes in the energy balance point. To better address this issue, calculations of the combustion heat for different fuels were conducted (Figures 13 and 14). The combustion heat for crude oil was determined when its energy balance concentration matched that of diesel, as well as when the combustion heat for crude oil was equivalent to that of diesel with the same cost.

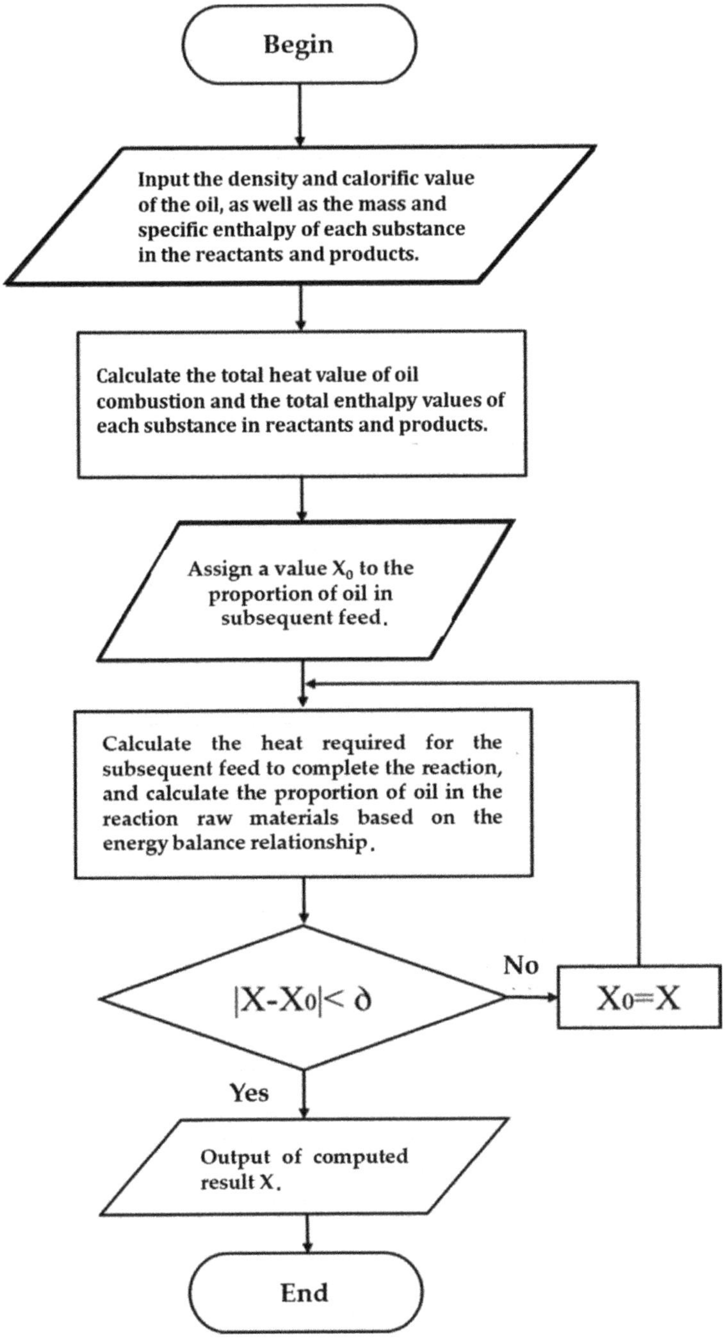

Figure 11. Flow for calculating energy balance concentration of supercritical water-diesel/crude oil reaction.

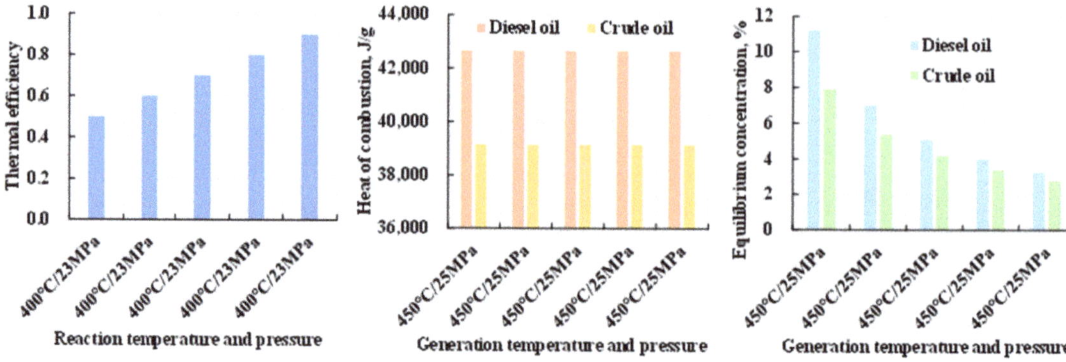

Figure 12. Energy balance concentration of supercritical water–diesel/crude oil reaction.

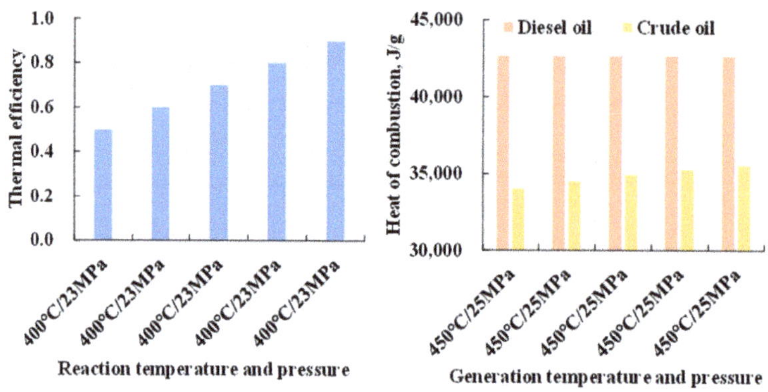

Figure 13. Heat of combustion in supercritical water–diesel/crude oil reaction with the same crude oil equilibrium concentration and diesel equilibrium concentration.

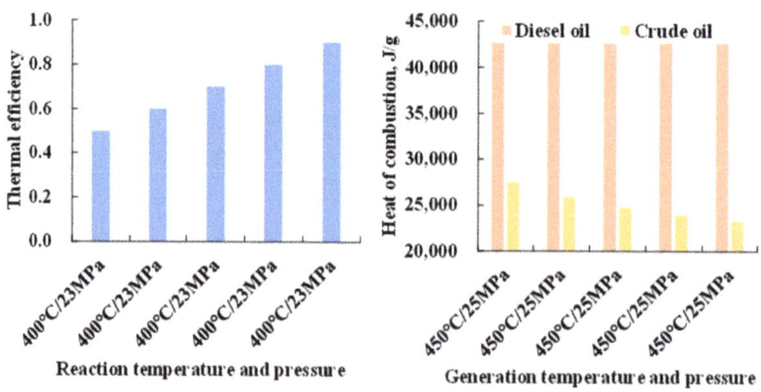

Figure 14. Heat of combustion in supercritical water–diesel/crude oil reaction with the same cost of crude oil diesel.

5. Feasibility Analysis of Crude Oil Replacing Diesel Oil to Prepare Supercritical Multicomponent Thermal Fluid

5.1. Reaction Mechanism Feasibility

From the various research results mentioned earlier, it is evident that the mechanisms of the reactions between supercritical water and diesel, as well as supercritical water and crude oil, are generally similar and can be divided into two stages. In the first stage, supercritical water disperses the continuous diesel/crude oil, and in the second stage, under the combined action of supercritical water and oxygen, the oil undergoes complete oxidation. Intermediate products are generated during the oxidation process. However, due to the presence of more long-chain compounds in crude oil, during the first stage of the reaction with supercritical water, there is not only physical dispersion but also a noticeable cracking process. This is the key distinction between the mechanism of the reaction between supercritical water and diesel, and the one between supercritical water and crude oil.

Therefore, it can be considered that utilizing crude oil as a substitute for diesel in the preparation of supercritical multicomponent thermal fluids is feasible and viable at a mechanistic level.

5.2. Economic Feasibility

Based on the current international oil prices and domestic refined oil prices, it is apparent that the price of 0# diesel is 8.13 yuan per liter, while the international Brent crude oil is equivalent to approximately 3.92 yuan per liter. This implies that the unit price of diesel is roughly twice that of crude oil. However, as calculated in the previous section using the energy balance point, under the same generator efficiency, the energy balance point for crude oil is lower. As long as the combustion heat of crude oil is not lower than 27,000 J/g and the generator efficiency is above 50%, it is economically feasible. Therefore, solely from the perspective of fuel consumption, the use of crude oil demonstrates favorable economic feasibility.

5.3. Technical Feasibility

From the aforementioned research results, it is evident that regardless of whether it is the supercritical water–diesel reaction or the supercritical water–crude oil reaction, there will be coke as an intermediate product. Moreover, notably, the amount of coke in the intermediate products of the supercritical water–crude oil reaction is higher than that in the supercritical water–diesel reaction. Additionally, experimental tests indicate that the oxygen consumption of diesel is higher than that of crude oil.

In the actual preparation of supercritical multicomponent thermal fluids in oil fields, when using diesel, more air needs to be injected into the generator, resulting in a higher flow rate and a further reduction in the likelihood of coke generation and deposition in the generator. However, for crude oil, the possibility of coke generation and deposition in the generator exists, especially when using high carbon–hydrogen ratio crude oil (i.e., high viscosity crude oil).

In summary, when preparing supercritical multicomponent thermal fluids using crude oil, at the technical level, attention should be paid to inject a sufficient amount of air into the generator. This is to ensure that the crude oil reacts completely, thereby reducing the generation of coke.

6. Conclusions

(1) In reactions involving supercritical water with crude oil or diesel, an increased proportion of water among the reactants results in a greater share of supercritical water in the resulting products. This, in turn, leads to an elevated specific enthalpy within the thermal fluid. The total enthalpy of the supercritical multicomponent thermal fluid produced from the interaction between diesel and supercritical water has an average value of 180,934.940 J, which exceeds the total enthalpy of the fluid produced from the reaction with crude oil, with an average value of 168,630.034 J. Conversely, the specific enthalpies of these two scenarios are inversely related.

(2) The combustion heat of crude oil is close to that of diesel, but when the generator efficiency is the same, the energy balance concentration for crude oil is lower than that for diesel. In other words, if the crude oil is used to generate supercritical

multicomponent thermal fluids in oil fields, the required concentration would be lower than that for diesel, meaning less quantity would be needed.

(3) Considering the reaction mechanisms, economic benefits, and technical conditions, using crude oil instead of diesel to generate supercritical multicomponent thermal fluids is feasible and has favorable potential.

Author Contributions: J.T., Q.F., Y.L., H.J. and Z.Q. conceived and designed the experiments; J.T. conducted the experiments and analyzed the data; J.T., Q.F. and S.Y. wrote the paper. All authors have read and agreed to the published version of this manuscript.

Funding: This research was funded by the National Natural Science Foundation of China (grant number 52004048, U22B2074) and the Natural Science Foundation of Chongqing Municipality, China (grant number cstc2020jcyj msxmX0856).

Data Availability Statement: The data that support the findings of this study are available from the corresponding author.

Conflicts of Interest: The authors declare no conflicts of interest.

References

1. Yu, L.D. Distribution of world heavy oil reserves and its recovery technologies and future. *Spec. Oil Gas Reserv.* **2001**, *8*, 98–103.
2. Dong, X.; Liu, H.; Chen, Z.; Wu, K.; Lu, N.; Zhang, Q. Enhanced oil recovery techniques for heavy oil and oil sands reservoirs after steam injection. *Appl. Energy* **2019**, *239*, 1190–1211. [CrossRef]
3. Lu, C.; Liu, H.; Zhao, W.; Lu, K.; Liu, Y.; Tian, J.; Tan, X. Experimental Investigation of In-Situ Emulsion Formation To Improve Viscous-Oil Recovery in Steam-Injection Process Assisted by Viscosity Reducer. *SPE J.* **2017**, *22*, 130–137. [CrossRef]
4. Zhang, Z.D.; Xu, J.; Li, J.H. *Gas Huff and Puff Technology for Enhanced Recovery in Deep Heavy Oil Reservoirs*; Petroleum Industry Press: Beijing, China, 2014.
5. Lyu, X.; Liu, H.; Pang, Z.; Sun, Z. Visualized study of thermochemistry assisted steam flooding to improve oil recovery in heavy oil reservoir with glass micromodels. *Fuel* **2018**, *218*, 118–126. [CrossRef]
6. Zhou, S.W.; Guo, L.J.; Li, Q.P. Multi-source Multi- thermal Fluid Generation System and Method. CN Patent CN106640007A, 10 May 2017.
7. Xu, J.L.; Kou, J.J.; Guo, L.J. Experimental study on oil-containing wastewater gasification in supercritical water in a continuous system. *Int. J. Hydrogen Energy* **2019**, *44*, 15871–15881. [CrossRef]
8. Sun, X.F.; Cai, J.M.; Li, X.Y. Experimental investigation of a novel method for heavy oil recovery using supercritical multithermal fluid flooding. *Appl. Therm. Eng.* **2020**, *185*, 116–130. [CrossRef]
9. Huang, Z.J.; Zhao, Q.Y.; Chen, L.; Miao, Y.; Wang, Y.C.; Jin, H.; Guo, L.J. Fundamentals of Enhanced Heavy Oil Recovery by Supercritical Multi-Component Thermal Fluid Flooding. *J. Eng. Thermophys.* **2022**, *43*, 974–981.
10. Tian, J.; Yan, W.; Qi, Z.; Huang, S.; Yuan, Y.; Dong, M. Cyclic Supercritical Multi-Thermal Fluid Stimulation Process: A Novel Improved-Oil-Recovery Technique for Offshore Heavy Oil Reservoir. *Energies* **2022**, *15*, 9189. [CrossRef]
11. Liu, Z.Y.; Li, Y.H.; Xu, L. Supercritical hydrothermal upgrading of oil sand: A review. *Chem. Ind. Eng. Prog.* **2018**, *37*, 4606–4615.
12. Ramazan, O.C.; Salim, S.S. Upgrading blends of microalgae feedstocks and heavy oils in supercritical water. *J. Supercrit. Fluids* **2018**, *133*, 674–682.
13. Li, Y.; Wang, S. Supercritical water oxidation for environmentally friendly treatment of organic wastes. In *Advanced Supercritical Fluids Technologies*; IntechOpen: Rijeka, Croatia, 2019.
14. Liu, J.; Hu, N.; Fan, L.W. Simulation on the hydrogen oxidation reactor for supercritical H2O/CO2 working fluid. *J. Chem. Eng. Chin. Univ.* **2021**, *35*, 298–306.
15. Qin, Q.; Xia, X.B.; Li, S.B. Supercritical water oxidation and its application in radioactive waste treatment. *Ind. Water Treat.* **2022**, *42*, 38–45.
16. Liao, W.; Zhu, T.F.; Liao, C.H. Application of Supercritical Water Oxidation Technology in Energy Transformation. *Technol. Water Treat.* **2019**, *45*, 14–17.
17. Cocero, M.J.; Alonso, E.; Sanz, M.T.; Fdz-Polanco, F. Supercritical water oxidation process under energetically self-sufficient operation. *J. Supercrit. Fluids* **2004**, *24*, 37–46. [CrossRef]
18. Qi, W.; Yong, K.L.; Rong, Z. Supercritical Water Oxidation Kinetics of Textiles Dyeing Wastewater. *Adv. Mater. Res.* **2013**, *634–638*, 302–306.
19. Zan, Y.F.; Wang, S.Z.; Zhang, Q.M. Experimental Studies on the Supercritical Water Oxidation of Municipal Sludge and its Reaction Heat. *J. Chem. Eng. Chin. Univ.* **2006**, *20*, 379–384.

20. Wang, Y.Z.; Yu, H.; Sheng, J.P. Treatment of biochemical sludge from coal gasification wastewater by supercritical water oxidation process. *Chem. Eng.* **2015**, *43*, 11–15.
21. Xia, Q.Y.; Guo, W.M.; Shen, Z.M. Research on Supercritical Water Oxidation of Industrial Wastewater. *Saf. Environ. Eng.* **2014**, *21*, 78–83.

Disclaimer/Publisher's Note: The statements, opinions and data contained in all publications are solely those of the individual author(s) and contributor(s) and not of MDPI and/or the editor(s). MDPI and/or the editor(s) disclaim responsibility for any injury to people or property resulting from any ideas, methods, instructions or products referred to in the content.

energies

Article

Composition and Injection Rate Co-Optimization Method of Supercritical Multicomponent Thermal Fluid Used for Offshore Heavy Oil Thermal Recovery

Shenyao Yang [1,2], Zhilin Qi [1], Jie Tian [1], Mingda Dong [1], Wei Zhang [1,*] and Wende Yan [1,*]

[1] The Key Laboratory of Unconventional Oil and Gas Green and Efficient Development of Chongqing Municipality, Chongqing University of Science and Technology, Chongqing 401331, China; yangshenyaocq@163.com (S.Y.)

[2] The Key Laboratory of Enhanced Oil and Gas Recovery of Education Ministry, Northeast Petroleum University, Daqing 163000, China

* Correspondence: zhangwei2021300@163.com (W.Z.); yanwende@cqust.edu.cn (W.Y.)

Abstract: Supercritical multicomponent thermal fluid injection is a new technology with great potential for offshore heavy oil thermal recovery. In the process of thermal fluid generation, the reaction conditions including temperature, pressure, and the organic mass concentration in the reaction material will significantly affect its composition and injection rate and will further affect the thermal recovery and development quality of heavy oil. However, there is a lack of relevant research on the variation rules and control methods of the composition and injection rate of supercritical multicomponent thermal fluids, resulting in a lack of technical mechanisms for effective optimization. To fill this gap, a reaction molecular dynamics simulation method was used to simulate thermal fluid generation under different temperatures, pressures, and organic mass concentrations. The changes in thermal fluid composition and yield with reaction conditions were studied, and a control model of thermal fluid composition and yield was established. The proportional relationship between the thermal fluid generation scale of an offshore heavy oil platform and the simulated thermal fluid generation scale is analyzed, and a collaborative optimization method of thermal fluid composition and injection rate in field applications is proposed. The results show the following: (1) The higher the mass concentration of organic matter, the higher the content of supercritical carbon dioxide and supercritical nitrogen in thermal fluids, and the lower the content of supercritical water. (2) The higher the temperature and pressure, the higher the thermal fluid yield, and the higher the organic mass concentration, the lower the thermal fluid yield. (3) The component fitting model conforms to the power function relationship, and the coefficient of determination R^2 is greater than 0.9; the yield fitting model conforms to the modified inverse linear logarithmic function relationship, the determination coefficient R^2 is greater than 0.8, and the fitting degree is high. (4) The ratio between the actual injection rate of thermal fluids in the mine field and the molecular simulated thermal fluid yield is the ratio of organic matter mass in the platform thermal fluid generator and organic matter mass in the simulated box. (5) Based on the composition and yield control model, combined with the simulation of the ratio relationship between yield and injection rate in the field, a collaborative optimization method of thermal fluid composition and injection rate was established. The research results can provide an effective technical method for predicting, controlling, and optimizing the composition and injection rate of supercritical multicomponent thermal fluids.

Keywords: supercritical multicomponent thermal fluids; offshore heavy oil thermal recovery; supercritical multicomponent thermal fluid composition; supercritical multicomponent thermal fluid yield; co-optimization; molecular dynamics simulation

1. Introduction

Supercritical multicomponent thermal fluids (scMCTF) are an innovative heat carrier used for thermal recovery of offshore heavy oil. They consist of supercritical water (scH_2O), supercritical carbon dioxide (scCO_2), and supercritical nitrogen (scN_2), offering superior heat-carrying, oil-solubilizing, viscosity-reducing, and pressure-maintaining properties. Compared to traditional steam and conventional multicomponent thermal fluids, scMCTFs significantly enhance the recovery rate of heavy oil [1–4]. Additionally, waste fluids produced from oil wells can be used as feedstock to generate scMCTFs, allowing the generation system to be self-sustaining without external energy inputs [5,6]. Due to its notable advantages in enhancing recovery efficiency and reducing costs, scMCTF injection is widely regarded as one of the most promising and valuable next-generation thermal recovery technologies for offshore heavy oil. It also has potential applications in onshore heavy oil fields [7,8].

The typical principle behind scMCTF generation involves gasification and oxidation reactions of organic matter in a supercritical water environment [9]. Reaction conditions influence the composition and yield of scMCTFs, which in turn determine the physical properties of the thermal fluid and the rate at which the scMCTF is injected from offshore platforms into reservoirs. The properties and injection rate of scMCTFs are key factors affecting the effectiveness of thermal recovery [10,11]. Therefore, it is critical to study how scMCTF composition and yield vary with reaction conditions and to establish a control model that enables precise adjustment of scMCTF composition and yield. This will help optimize the parameters for scMCTF injection, improving the quality of offshore heavy oil recovery.

Currently, research on scMCTF injection for heavy oil recovery is scarce, with related studies mainly emerging in the past three years [1–9,12]. No reports have been found on control models or optimization methods for scMCTF composition and yield, highlighting the lack of effective technical approaches. In this context, molecular dynamics simulations are employed to study scMCTF composition and yield under varying temperatures, pressures, and feedstock concentrations. The goal is to establish a control model for scMCTF composition and yield, analyze the relationship between the scale of scMCTF generation in offshore platforms and simulation results, and propose methods to optimize both the composition and injection rate of scMCTFs in field applications. These research findings aim to provide effective technical methods for predicting, controlling, and optimizing scMCTF composition and injection rates, offering crucial technical support for parameter optimization in future offshore heavy oil development projects.

2. Reaction Molecular Dynamics Simulation Model

2.1. Modeling System

Based on the Materials Studio platform for atom or molecule system construction, optimization, and simulation, a multicomponent diesel molecular model was constructed using paraffins, cycloalkanes, monocyclic aromatics, and bicyclic aromatics as model compounds [13], and the ratio of paraffins, cycloalkanes, monocyclic aromatics, and bicyclic aromatics molecules number is 5:3:1:1.

Subsequently, after fixing the number of organic molecules, different numbers of water molecules were added to establish reaction systems with varying organic concentrations, followed by molecular geometry optimization. The organic concentration in the system ranged from 8.39% to 25.14% (Figure 1). Reaction systems with higher organic concentrations have fewer water molecules, fewer total molecules, and looser molecular arrangements.

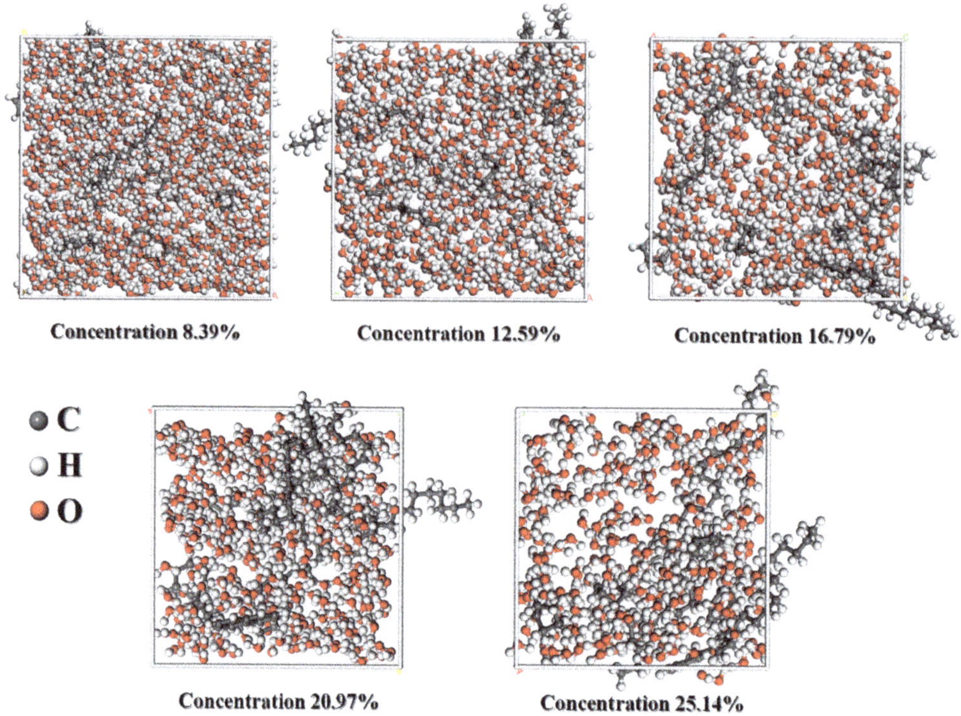

Figure 1. Concentration reaction system.

2.2. Simulated Force Field and Program Setup

The reactive molecular dynamics simulations were carried out using the LAMMPS (Large-scale Atomic or Molecular Massively Parallel Simulator) platform. The ReaxFF reactive force field applied was the C/H/O force field proposed by Chowdhury and van Duin in 2017 [14], which has been widely utilized for simulating the combustion, pyrolysis, and gasification of organic compounds under supercritical water conditions [15–17]. The global simulation time step was set to 0.1 fs, and the simulation procedure included the following steps: ① Charge equilibration using the QEQ method. ② System energy minimization via the conjugate gradient method. ③ Random initialization of atomic velocities at 200 °C according to the Maxwell–Boltzmann distribution. ④ System relaxation under specified temperature and pressure based on the NPT ensemble using the Nose–Hoover method to control temperature and pressure, with damping coefficients 100 times and 1000 times the time step for temperature and pressure, respectively. The relaxation time was 10 ps, allowing the system to reach equilibrium. ⑤ Resetting of the time step after removing the ensemble setting. ⑥ Reaction simulation of supercritical multicomponent thermal fluids using the NPT ensemble, with the reaction temperature and pressure set according to simulation settings. The damping coefficients for temperature and pressure remained unchanged. ⑦ Output and recording of system parameters such as temperature, pressure, system volume, and atomic positions every 100 time steps, along with the storage of product information.

2.3. Model Verification

Based on the different organic matter concentration reaction systems established, the molecular simulation software LAMMPS 64-bit 2Aug2023-MSMPI was used to calculate the system's density. Additionally, molecular simulations of supercritical multicomponent thermal fluid (scMCTF) generation were performed under varying temperatures

(3500–5500 °C) and pressures (23–35 MPa). Physical experiments were also conducted under real laboratory conditions for systems with the same organic matter concentrations, covering temperatures (400–500 °C) and pressures (23–25 MPa).

In reactive molecular dynamics simulations, reaction temperatures are typically elevated to accelerate reaction rates because the simulation timescales are often on the order of picoseconds (ps) to nanoseconds (ns) [18,19]. According to the Arrhenius equation describing the relationship between reaction temperature and reaction rate [20], scholars such as Voter and SΦrensen introduced the Temperature-Accelerated Dynamics (TAD) method to increase the simulation temperature, which speeds up the reaction process without significantly affecting the reaction mechanisms [21]. This method has been widely applied and validated [22–24]. Therefore, the simulation temperatures in this study were set higher than those in the physical experiments.

A comparison of the simulation and experimental results (Figure 2) shows that the densities of the simulated and measured reaction systems were similar. Furthermore, the mass concentrations of products generated under different reaction pressures, temperatures, and organic matter concentrations from the simulations were numerically close to those obtained from the physical experiments, with highly consistent trends. Therefore, it can be concluded that reactive molecular dynamics simulations can accurately reflect the actual scMCTF generation process.

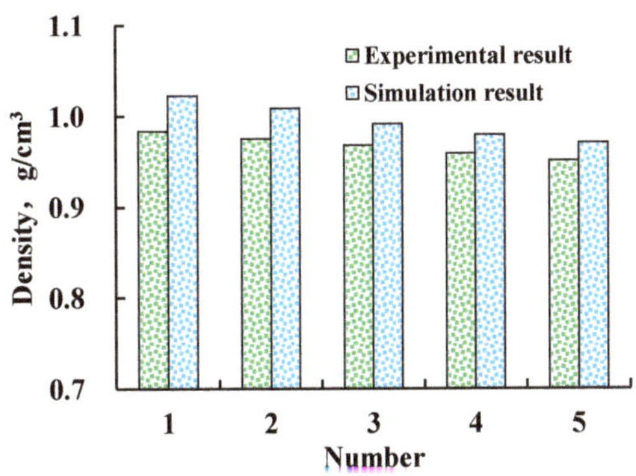

(**a**) Comparison of reaction system density

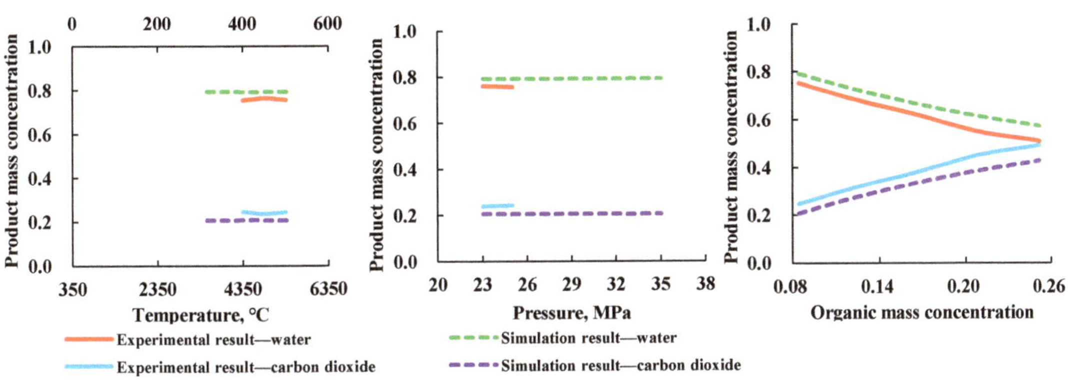

(**b**) Comparison of product mass concentration under different reaction conditions

Figure 2. Comparison of molecular dynamics simulation results and experimental results.

3. Change Law of scMCTF Composition and Yield

Based on the established reactive molecular dynamics model, a simulation study of the supercritical multicomponent thermal fluid (scMCTF) generation process was conducted. The simulation parameters were set as follows: a temperature range of 3500 °C to 5500 °C with a step size of 500 °C; a pressure range of 23 MPa to 35 MPa with a step size of 3 MPa; and an organic matter concentration range of 8.39% to 25.14%, with a step size of 4.20%. In total, the reactive molecular dynamics simulations included 125 sets of conditions (5 × 5 × 5).

3.1. Change Law of Composition

Based on the mass concentrations of each component in scMCTF generated under different temperatures, pressures, and organic matter concentrations in the feedstock (Figure 3), the component concentration–temperature–pressure surface for different organic matter concentrations is essentially flat. This indicates that the composition of scMCTF does not significantly vary with temperature and pressure, meaning that temperature and pressure are not the primary controlling factors for scMCTF composition. However, the concentration of organic matter in the feedstock has a greater impact on the composition of scMCTF. As shown in Figure 3, the concentrations of $scCO_2$ and scN_2 increase with the rise in organic matter concentration, while the concentration of scH_2O increases as the organic matter concentration decreases.

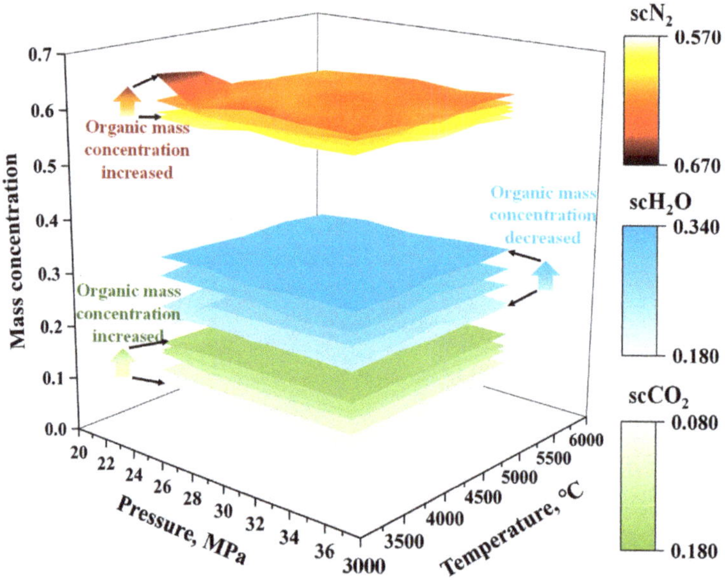

Figure 3. Change law of scMCTF composition.

This is because higher organic matter concentrations in the feedstock mean more organic content, leading to a higher total carbon count in the reaction system, which ultimately results in the formation of more $scCO_2$ after the reaction. At the same time, the more organic matter there is, the more feedwater is consumed in the reaction, so the concentration of scH_2O decreases as the organic matter concentration increases. Correspondingly, since the generation of $scCO_2$ and scH_2O consumes oxygen from the air, the more that is produced, the more oxygen is used. This means that the remaining nitrogen in the air, which does not participate in the reaction, increases, leading to a rise in scN_2 concentration with increasing organic matter concentration.

3.2. Change Law of Yield

Based on the scMCTF yield generated under different temperatures, pressures, and organic matter concentrations in the feedstock (Figure 4), it can be seen that higher pressure results in a higher thermal fluid yield, although this effect is relatively small. As the organic matter concentration in the feedstock increases, the yield decreases, while higher reaction temperatures lead to a higher yield, with temperature having a very significant impact on yield.

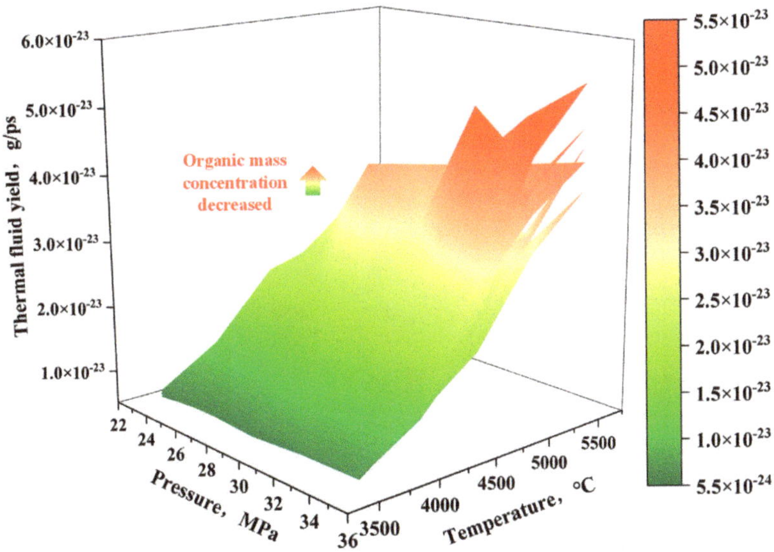

Figure 4. Change law of thermal fluids yield.

From the perspective of molecular collision theory, chemical reactions occur through collisions between molecules. The higher the pressure, the closer the initial distance between molecules, increasing the probability of collisions and thus raising the reaction yield as pressure increases. The effect of organic matter concentration on yield is relative. A higher organic matter concentration means more molecules per unit volume, which favors collisions and accelerates the reaction rate. However, the gasification reaction of organic matter in a supercritical water environment is sensitive to water concentration; the lower the water concentration and the higher the organic matter concentration, the less active the gasification reaction becomes [25]. As a result, the influence of organic matter concentration on yield is a combined effect of these two factors, with the latter being dominant. Higher temperatures increase molecular speed and kinetic energy, raising both collision energy and collision probability, thus enhancing the reaction rate and ultimately increasing the yield.

4. Model of scMCTF Composition and Yield Control

4.1. Model of Composition Control

Based on the previous analysis, it is evident that temperature and pressure have minimal impact on the composition of scMCTF. Therefore, their influence is neglected in the modeling process, and a correlation model is established between the scMCTF composition and the organic matter concentration in the feedstock. Additionally, since the sum of the three components in scMCTF equals 1, only the models for $scCO_2$ and scH_2O with respect to the organic matter concentration are developed.

As seen from the fitting results of $scCO_2$ and scH_2O with organic matter concentration (Figure 5), both $scCO_2$ and scH_2O follow a power function relationship with organic matter concentration, with the fitting coefficients of determination (R^2) exceeding 0.99, indicating a high degree of fit. Therefore, an scMCTF composition control model can be established based on these fitting equations (Equations (1)–(3)), and the organic matter concentration required for the reaction can be calculated based on the desired content of a particular scMCTF component, thereby achieving control over the composition of scMCTF injected into the reservoir.

$$C(scH_2O) = 0.1360C(OMC)^{-0.3582} \tag{1}$$

$$C(scCO_2) = 0.3924C(OMC)^{0.6126} \tag{2}$$

$$C(scN_2) = 1 - C(scH_2O) - C(scCO_2) \tag{3}$$

where $C(scH_2O)$, $C(scCO_2)$, and $C(scN_2)$ represent the mass concentrations of supercritical water, supercritical carbon dioxide, and supercritical nitrogen, respectively; f. OMC is the mass concentration of organic matter in the feedstock, f.

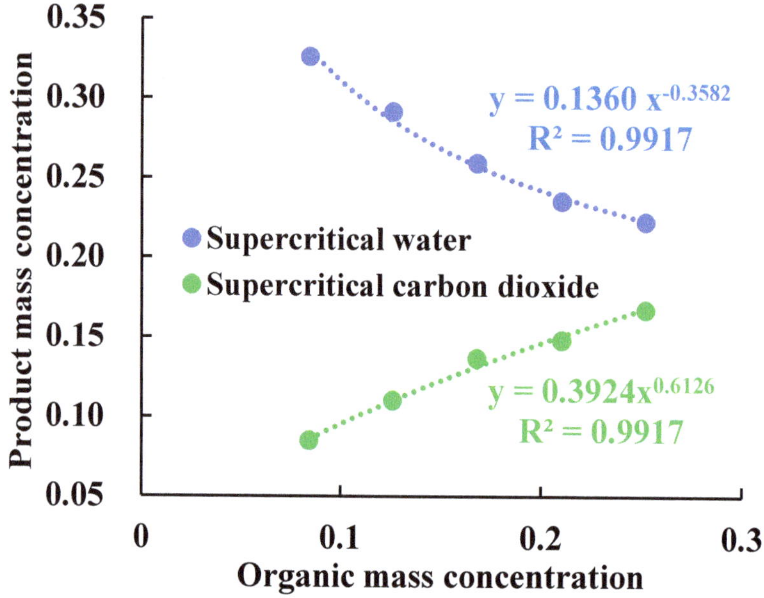

Figure 5. Relationship model between content and mass concentration.

4.2. Model of Yield Control

Compared to the composition control model, the yield control model is more complex because it is influenced by three factors: temperature, pressure, and organic matter concentration. Therefore, a control model is established based on a quadratic fitting approach. First, temperature and pressure are treated as independent variables, and a modified inverse linear logarithmic model is used to develop the relationship between temperature, pressure, and thermal fluid yield under different organic matter concentration conditions (Equation (4), Table 1). It is worth noting that 10^{-24} in Equation (4) is just an order of magnitude conversion coefficient of the fitting results and has no actual physical significance.

$$v = 10^{-24}\frac{TP}{a + b\ln(T)} \tag{4}$$

where v represents the molecular simulation thermal fluid yield, g·ps^{-1}. T is the reaction temperature, °C. P is the reaction pressure, and MPa. a and b are the model fitting coefficients, which are dimensionless.

Table 1. Model coefficients of different organic matter mass concentration conditions.

Number	Organic Mass Concentration, f	Fitting Coefficient *a*	Fitting Coefficient *b*	R^2
1	0.0839	1.11×10^5	-1.22×10^4	0.87
2	0.1259	1.08×10^5	-1.23×10^4	0.80
3	0.1679	1.37×10^5	-1.47×10^4	0.90
4	0.2097	1.37×10^5	-1.54×10^4	0.88
5	0.2514	1.57×10^5	-1.75×10^4	0.92

Based on the fitting coefficients *a*, *b*, and the corresponding organic matter concentration data, the relationship curves of fitting coefficient *a*—organic matter concentration and fitting coefficient *b*—organic matter concentration were plotted by Microsoft Office EXCEL 2019 (Figure 6). As shown in Figure 6, the fitting coefficient *a* increases as the feedstock organic matter concentration increases, while fitting coefficient *b* decreases with the increase of the feedstock organic matter concentration. The fitting equation for *a*, organic matter concentration, has a coefficient of determination R^2 of 0.88, and the fitting equation for *b*, organic matter concentration, has a coefficient of determination R^2 of 0.94. Both show a good linear relationship with the organic matter concentration. Therefore, the relationship models for fitting coefficient *a*, organic matter concentration, and fitting coefficient *b*, organic matter concentration, can be established (Equations (5) and (6)).

$$a = 2.90 \times 10^5 \mathrm{OMC} + 8.14 \times 10^4 \quad (5)$$

$$b = -3.27 \times 10^4 \mathrm{OMC} - 8.95 \times 10^3 \quad (6)$$

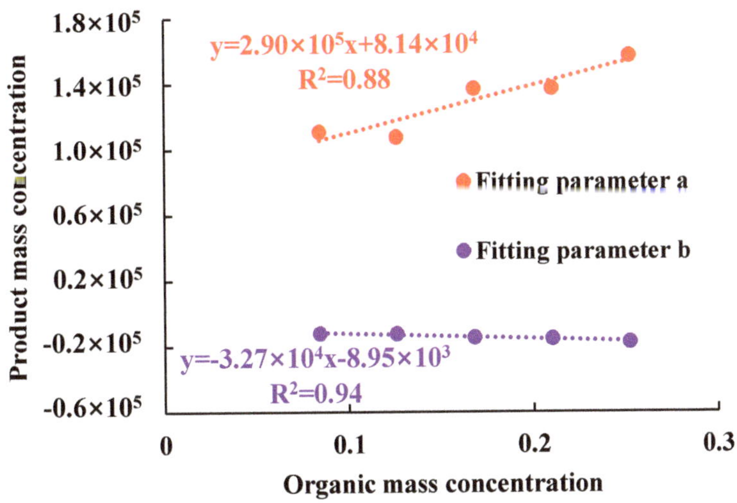

Figure 6. Relationship model between fitting coefficient and mass concentration.

By combining the temperature–pressure–thermal fluid yield correlation model (Equation (4)) with the fitting coefficient–organic matter concentration correlation models (Equations (5) and (6)), an organic matter concentration–temperature–pressure–thermal fluid yield correlation model is established, namely the scMCTF yield control model (Equation (7)). Using Equation (7), the scMCTF yield under specific temperature, pressure, and feedstock organic matter concentration conditions can be calculated, thereby achieving the control of the scMCTF yield.

$$v = \frac{10^{-24}TP}{(2.90 \times 10^5\text{OMC} + 8.14 \times 10^4) - (3.27 \times 10^4\text{OMC} + 8.95 \times 10^3)\ln(T)} \quad (7)$$

5. Optimization Method of Synergistic scMCTF Composition and Yield

5.1. Conversion Relationship Between Yield and Injection Rate

In practical field applications, the injection rate of thermal fluids is typically above the ton/day scale [26], which is far greater than the scMCTF yield calculated using the reactive molecular dynamics model. This difference arises from the scale discrepancy between the thermal fluid generation systems. On-site systems are large-scale, meter-sized scMCTF generators, whereas the molecular model represents a nanoscale simulation box. Therefore, it is necessary to analyze the relationship between these two scenarios and establish a corresponding conversion model.

The scMCTF generation principle in large generators is consistent with the principle used in the molecular dynamic simulation model. Based on this, when the temperature and pressure inside the generator are the same as those inside the simulation box, the simulation box can be considered a microelement within the generator (Figure 7). This is because, under the same temperature and pressure conditions, substances share the same thermodynamic state.

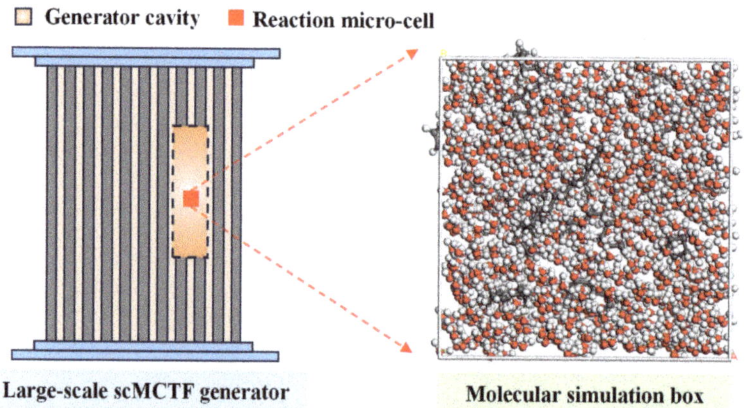

Figure 7. Comparison of scale relationship between scMCTF generator and molecular simulation box.

Therefore, the scMCTF generator in the field can be viewed as being composed of countless microelements or, in other words, countless simulation boxes. The conversion relationship between them depends on the number of microelements inside the generator. The scMCTF generation scale in the field is X times the scMCTF generation scale in the simulation box, where X is the ratio of the generator's internal volume to the simulation box volume. Under the same temperature and pressure conditions, the fluid density inside the generator and the simulation box is identical, meaning the volume ratio X equals the mass ratio Y of the fluid inside them. Furthermore, under the same organic matter concentration, the mass ratio of the fluid is effectively equal to the mass ratio of the organic matter, Z. For example, if the mass of organic matter in the simulation box is 1 g and the mass of organic matter in the field generator is 1000 g, the scMCTF injection rate in the field will be 1000 times the scMCTF yield in the simulation box. In the simulations mentioned earlier, the organic matter used was a mixture of five molecules of hexadecane (paraffin), three molecules of cyclopentane (naphthene), one molecule of benzene (monocyclic aromatic hydrocarbon), and one molecule of naphthalene (bicyclic aromatic hydrocarbon), with a total mass of 2.572×10^{-24} kg. Based on this, a conversion relationship (Equation (8)) between the scMCTF yield and injection rate is established. Combining Equations (7) and (8), the scMCTF injection rate for offshore heavy oil development platforms under various reaction conditions can be calculated.

$$I = \frac{m}{2.572 \times 10^{-24}} v \tag{8}$$

where I represents the scMCTF injection rate for offshore heavy oil development platforms, g·ps^{-1}. m represents the mass of organic matter placed in the platform's scMCTF generator, kg.

5.2. Synergistic Optimization Method

As previously discussed, in field applications, the composition of the scMCTF injected into the heavy oil reservoir and the injection rate are two crucial parameters. These parameters are determined simultaneously at the scMCTF generation point on the platform and are not affected in subsequent flow stages. Therefore, it is essential to optimize both the composition and the injection rate in coordination.

Based on the established control models (Equations (1)–(3), (7) and (8)), the composition of scMCTF is primarily controlled by the organic matter concentration, while the injection rate of scMCTF is influenced by temperature, pressure, and organic matter concentration. Thus, a method for optimizing both the scMCTF composition and injection rate is proposed (Figure 8). First, determine the required organic matter concentration based on the target scMCTF composition. Then, based on the temperature and pressure tolerance limits of the platform's scMCTF generator, along with other technical and economic factors, define an acceptable range for reaction temperature and pressure. Within this range, select a specific temperature–pressure combination. Next, input the determined organic matter concentration and selected temperature–pressure data into the scMCTF yield control equation to calculate the corresponding yield under these conditions. Using the conversion relationship between the yield and injection rate, determine the scMCTF injection rate. Finally, assess whether the obtained injection rate meets the target. If it does, the optimization process ends; if not, adjust the temperature and pressure values and repeat the optimization process until the injection rate reaches the desired target.

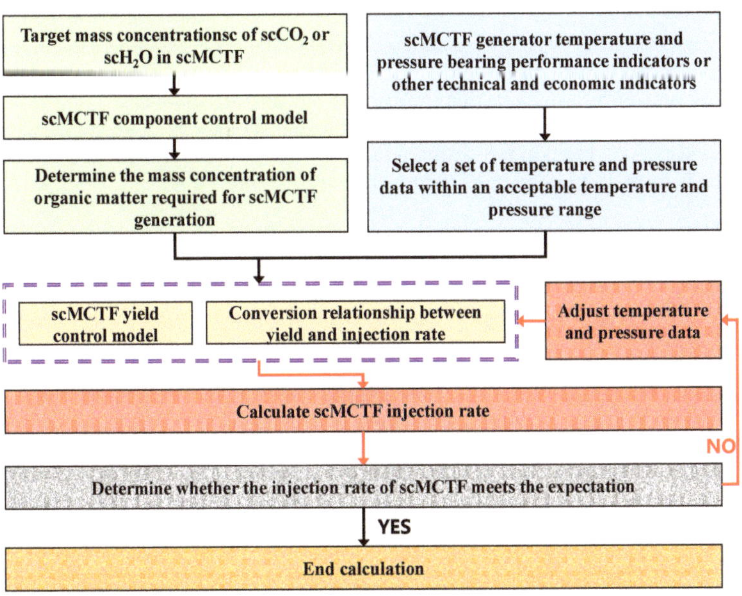

Figure 8. Work flow.

6. Discussion

Currently, laboratory studies on steam-assisted gravity drainage for heavy oil recovery using injected scMCTF rarely consider the impact of differences in fluid composition or injection rate on recovery efficiency [1–3,27]. However, optimizing the content of non-condensable gasses in the thermal fluid and the injection rate is critical for adjusting injection parameters to achieve the best thermal recovery results [28]. The newly proposed synergistic optimization method, based on the scMCTF composition and yield control model, offers a novel global optimization approach. It enables the simultaneous calculation of optimal reaction conditions required to achieve the desired scMCTF composition and injection rate while considering technical, economic, and equipment limitations on-site, thereby significantly enhancing injection operation efficiency.

Despite its advantages, the proposed method also has some limitations, which include the following two points:

(1) Inability to independently control scCO_2 and scN_2 content. Since the control equations for scH_2O (Equation (1)) and scCO_2 (Equation (2)) are both primarily influenced by organic matter concentration, once one of these components is determined, the other is passively fixed. According to the relationship between scCO_2, scH_2O, and scN_2 (Equation (3)), once the scCO_2 and scH_2O contents are determined, the scN_2 content is also fixed. In other words, the control equations can only adjust the content of one of the non-condensable gasses, whereas current research often aims to determine the ratio of these two gasses [28].

(2) Limited consideration of influencing factors. The current experiments, simulations, and models focus only on the three most fundamental factors: temperature, pressure, and organic matter concentration. However, other factors also affect the reaction, such as the type of feedstock [29,30], reaction time [31,32], catalyst type, and concentration [33,34]. Expanding the range of influencing factors would enhance the adaptability of the optimization method, enabling its application to a broader set of field conditions and unlocking more opportunities for technical parameter optimization in the field, thereby improving the efficiency of scMCTF thermal recovery operations.

Given these limitations, future research should focus on the following key areas:

(1) Improving the quality of control equations to enable independent control of individual component concentrations.

(2) Developing control equations that incorporate additional factors based on field requirements, enhancing the versatility of the optimization method. This would expand its application scope and provide more flexibility for field parameter adjustments, leading to more efficient scMCTF thermal recovery operations.

7. Conclusions

(1) The influence of temperature and pressure on the composition of supercritical multicomponent thermal fluid (scMCTF) is not significant, while the composition is largely influenced by the organic matter concentration in the feedstock. As the organic matter concentration increases, the content of supercritical carbon dioxide (scCO_2) and supercritical nitrogen (scN_2) rises, while the content of supercritical water (scH_2O) decreases.

(2) Temperature, pressure, and the organic matter concentration in the feedstock all have an impact on the scMCTF yield. Higher temperatures and pressures lead to a higher scMCTF yield, while a higher organic matter concentration results in a lower yield.

(3) Using data fitting methods, control models for scMCTF composition and yield were developed. The scMCTF composition fitting model follows a power function relationship, with a coefficient of determination (R^2) greater than 0.9. The scMCTF yield fitting model follows a modified inverse linear logarithmic function, with an R^2 greater than 0.8, indicating a high degree of fit for both the composition and yield control models.

(4) The ratio between scMCTF production in molecular simulations and the actual scMCTF injection rate in the field is proportional to the ratio of the mass of organic matter in the platform's scMCTF generator to that in the simulation box. This allows for the conversion of scMCTF yields from simulation-scale to actual field-scale injection rates.
(5) Based on the established composition and yield control models, along with the proposed conversion relationship between the scMCTF yield and injection rate, a synergistic optimization method for scMCTF composition and the injection rate was developed. This provides an effective technical approach for simultaneously adjusting and optimizing scMCTF composition and the injection rate in field operations.

Author Contributions: Methodology, M.D. and W.Y.; Investigation, J.T.; Writing—original draft, S.Y.; Writing—review & editing, W.Z.; Supervision, Z.Q. and W.Y. All authors have read and agreed to the published version of the manuscript.

Funding: This work was supported by the National Natural Science Foundation of China (Grant number 52004048, U22B2074, 52104025); the Natural Science Foundation of Chongqing Municipality, China (Grant number cstc2020jcyj msxmX0856); the Science and Technology Research Program of Chongqing Municipal Education Commission (KJQN202001508); the Natural Science Foundation of Chongqing, China (CSTB2022NSCQ-MSX0858); the Science and Technology Research Program of Chongqing Municipal Education Commission (Grant No. KJZD-M202301501); and the Talent Plan project of Chongqing Municipality (cstc2022ycjh-bgzxm0055).

Data Availability Statement: The data presented in this study are available on request from the corresponding author. The data are not publicly available due to privacy restrictions.

Conflicts of Interest: The authors declare no conflict of interest.

References

1. Li, X.; Sun, X.; Cai, J.; Zhang, Q.; Pan, X.; Zhang, Y. Experimental investigation on supercritical multi-thermal fluid flooding using a novel 2-dimensional model. *Energy* **2023**, *283*, 129136. [CrossRef]
2. Huang, Z.; Zhao, Q.; Chen, L.; Guo, L.; Miao, Y.; Wang, Y.; Jin, H. Experimental investigation of enhanced oil recovery and in-situ upgrading of heavy oil via CO_2- and N_2-assisted supercritical water flooding. *Chem. Eng. Sci.* **2023**, *268*, 118378. [CrossRef]
3. Huang, Z.; Zhao, Q.; Chen, L.; Miao, Y.; Wang, Y.; Jin, H.; Guo, L. Fundamentals of Enhanced Heavy Oil Recovery by Supercritical Multi-Component Thermal Fluid Flooding. *J. Eng. Thermophys.* **2022**, *43*, 974–981.
4. Cui, G.; Niu, Z.; Hu, Z.; Feng, X.; Chen, Z. The Production Analysis and Exploitation Scheme Design of a Special Offshore Heavy Oil Reservoir—First Offshore Artificial Island with Thermal Recovery. *J. Mar. Sci. Eng.* **2024**, *12*, 1186. [CrossRef]
5. Zhao, Q.Y.; Jin, H.; Xu, J.L.; Peng, Z.Y.; Wang, Y.C.; Guo, L.J. Theory and Technology of Multi-Component Supercritical Thermal Fluid Generation and In-Situ Conversion Technique for Heavy Crude Oil Resources Exploitation. *J. Xi'an Jiaotong Univ.* **2023**, *57*, 31–45.
6. Sun, X.; Li, X.; Tan, X.; Zheng, W.; Zhu, G.; Cai, J.; Zhang, Y. Pyrolysis of heavy oil in supercritical multi-thermal fluid: An effective recovery agent for heavy oils. *J. Pet. Sci. Eng.* **2021**, *196*, 107784. [CrossRef]
7. College of Petroleum and Natural Gas Engineering, Chongqing University of Science and Technology. The Key Project Launch and Program Review Meeting of the National Self-Science Fund Enterprise Innovation and Development Joint Fund of Our School Was Held. 2023. Available online: https://sgy.cqust.edu.cn/info/1221/2142.htm (accessed on 6 December 2023).
8. State Key Laboratory of Dynamic Engineering Multiphase Flow, Xi'an Jiaotong University. Academician Guo Liejin, Director of the Laboratory, and Young Teacher Zhao Qiuyang, Have Made Important Progress in the Research on the Mechanism of Supercritical Fluid Development of Unconventional Oil, and Their Results Have Been Selected as a Journal Cover Article. 2023. Available online: http://mfpe.xjtu.edu.cn/info/1077/6897.htm (accessed on 6 December 2023).
9. Xu, J.; Peng, Z.; Rong, S.; Zhao, Q.; Jin, H.; Guo, L.; Zhang, X. Optimal retrofit of a novel multi-component supercritical thermal fluid generation system via thermodynamic analysis. *Appl. Therm. Eng.* **2023**, *219*, 119511. [CrossRef]
10. Pang, Z.; Wu, Z.; Zhao, M. A novel method to calculate consumption of non-condensate gas during steam assistant gravity drainage in heavy oil reservoirs. *Energy* **2017**, *130*, 76–85. [CrossRef]
11. He, H.; Li, Q.; Zheng, H.; Liu, P.; Tang, J.; Ma, Y. Simulation and evaluation on enhanced oil recovery for steam huff and puff during the later phase in heavy oil Reservoir—A case study of block G in Liaohe oilfield, China. *J. Pet. Sci. Eng.* **2022**, *219*, 111092. [CrossRef]
12. Ilyushin, Y.; Afanaseva, O. Development of a spatial-distributed control system for preparation of pulse gas. *Int. Multidiscip. Sci. GeoConference SGEM* **2020**, *20*, 475–482.
13. Chen, Z.; Sun, W.; Zhao, L. Initial Mechanism and Kinetics of Diesel Incomplete Combustion: ReaxFF Molecular Dynamics Based on a Multicomponent Fuel Model. *J. Phys. Chem. C* **2019**, *123*, 8512–8521. [CrossRef]

14. Chowdhury, A.; van Duin, A.C.T. Extension of the ReaxFF Combustion Force Field toward Syngas Combustion and Initial Oxidation Kinetics. *J. Phys. Chem. A* **2017**, *121*, 1051–1068.
15. Lele, A.; Kwon, H.; Ganeshan, K.; Xuan, Y.; van Duin, A.C.T. ReaxFF molecular dynamics study on pyrolysis of bicyclic compounds for aviation fuel. *Fuel* **2021**, *297*, 120724. [CrossRef]
16. Li, H.; Xu, B.; Jin, H.; Luo, K.; Fan, J. Molecular dynamics investigation on the lignin gasification in supercritical water. *Fuel Process. Technol.* **2019**, *192*, 203–209. [CrossRef]
17. Zhang, X.; Liu, L.; Xiao, M.; Lang, B.; Xia, L.; Lin, R.; Wang, X. Insight into the thermal conversion of corn stalk gasification in supercritical water based on reactive molecular dynamics simulations. *J. Energy Inst.* **2023**, *106*, 101156. [CrossRef]
18. Liu, X.; Wang, T.; Chu, J.; He, M.; Li, Q.; Zhang, Y. Understanding lignin gasification in supercritical water using reactive molecular dynamics simulations. *Renew. Energy* **2020**, *161*, 858–866. [CrossRef]
19. Castro-Marcano, F.; Kamat, A.M.; Russo, M.F., Jr.; van Duin, A.C.T.; Mathews, J.P. Combustion of an Illinois No. 6 coal char simulated using an atomistic char representation and the ReaxFF reactive force field. *Combust. Flame* **2012**, *159*, 1272–1285. [CrossRef]
20. Dong, Y.; Wang, Y.; Zhang, F. *Inorganic and Analytical Chemistry*, 3rd ed.; Science Press: Beijing, China, 2010.
21. Sorensen, M.R.; Voter, A.F. Temperature-accelerated dynamics for simulation of infrequent events. *J. Chem. Phys.* **2000**, *112*, 9599–9606. [CrossRef]
22. Salmon, E.; van Duin, A.C.T.; Lorant, F.; Marquaire, P.M.; Goddard, W.A., III. Early maturation processes in coal. Part 2: Reactive dynamics simulations using the ReaxFF reactive force field on Morwell Brown coal structures. *Org. Geochem.* **2009**, *40*, 1195–1209. [CrossRef]
23. Zhao, J.S.; Huang, Z.Y.; Jin, G.F.; Gao, M.N.; Zhu, H.X. Reaction dynamics study of hypergolic bipropellants: Azide amine and dinitrogen Tetroxide. *Propellants Explos. Pyrotech.* **2021**, *46*, 1679–1686. [CrossRef]
24. Feng, M.; Jiang, X.Z.; Mao, Q.; Luo, K.H.; Hellier, P. Initiation mechanisms of enhanced pyrolysis and oxidation of JP-10 (exo-tetrahydrodicyclopentadiene) on functionalized graphene sheets: Insights from ReaxFF molecular dynamics simulations. *Fuel* **2019**, *254*, 115643. [CrossRef]
25. Guo, S.; Guo, L.; Cao, C.; Yin, J.; Lu, Y.; Zhang, X. Hydrogen production from glycerol by supercritical water gasification in a continuous flow tubular reactor. *Int. J. Hydrog. Energy* **2012**, *37*, 5559–5568. [CrossRef]
26. Jiang, W.D.; Luo, S.F.; Zhang, W.; Sun, Y.B.; Wang, S.H. Research and application of thermochemical huff and puff efficiency increasing technology for heavy oil in Bohai heavy oilfield. *China Offshore Oi Gas* **2024**, *36*, 141–148.
27. Li, X.Y.; Sun, X.F.; Cai, J.M.; Wang, X.Q.; Zhang, Q.Q.; Hu, J.P.; Wei, C.K. Physical simulation study on oil flooding characteristics of supercritical multiple-thermal fluids in offshore heavy oil reservoirs. *Fault-Block Oil Gas Field* **2023**, *30*, 545–551.
28. Sun, F.R.; Yao, Y.D.; Li, X.F.; Zhang, Y.; Ding, G.Y.; Zou, M. Evaluation of Heating Effect on the Horizontal Well in 35-2 Bohai Oilfield with Multiple Thermal Fluid Stimulation. *J. Beijing Inst. Petrochem. Technol.* **2017**, *25*, 5–8.
29. Susanti, R.F.; Dianningrum, L.W.; Yum, T.; Kim, Y.; Lee, Y.W.; Kim, J. High-yield hydrogen production by supercritical water gasification of various feedstocks: Alcohols, glucose, glycerol and long-chain alkanes. *Chem. Eng. Res. Des.* **2014**, *92*, 1834–1844. [CrossRef]
30. Zhang, H.; Wang, C.Y.; Chen, C.; Zhong, J. Study on the Reaction of Saturated Hydrocarbons of Oily Sludge in Supercritical Water. *Environ. Sci. Technol.* **2019**, *32*, 20–23,29.
31. Kou, J.; Feng, H.; Wei, W.; Wang, G.; Sun, J.; Jin, H.; Guo, L. Study on the detailed reaction pathway and catalytic mechanism of a Ni/ZrO_2 catalyst for supercritical water gasification of diesel oil. *Fuel* **2022**, *312*, 122849. [CrossRef]
32. Li, L.H.; Li, X.J.; Cao, W. Reaction pathway and kinetics study on supercritical water gasification of oily sludge. *J. Anal. Appl. Pyrolysis* **2023**, *170*, 105920. [CrossRef]
33. Kou, J.; Xu, J.; Jin, H.; Guo, L.; Zhang, D.; Cao, W. Evaluation of modified Ni/ZrO_2 catalysts for hydrogen production by supercritical water gasification of oil-containing wastewater. *Int. J. Hydrog. Energy* **2018**, *43*, 13896–13903. [CrossRef]
34. Peng, Z.; Rong, S.; Xu, J.; Luo, K.; Zhang, J.; Jin, H.; Guo, L. Hydrogen production from oilfield wastewater by gasification in supercritical water with a continuous system. *Fuel* **2023**, *344*, 128094. [CrossRef]

Disclaimer/Publisher's Note: The statements, opinions and data contained in all publications are solely those of the individual author(s) and contributor(s) and not of MDPI and/or the editor(s). MDPI and/or the editor(s) disclaim responsibility for any injury to people or property resulting from any ideas, methods, instructions or products referred to in the content.

energies

MDPI

Article

Influence of Reaction Conditions on the Yield of Supercritical Multicomponent Thermal Fluids

Wei Zhang [1], Zhilin Qi [1], Jie Tian [1], Fang Xu [2], Deyu Kong [2], Mingda Dong [1], Shenyao Yang [1,3,*] and Wende Yan [1,*]

1 The Key Laboratory of Unconventional Oil and Gas Green and Efficient Development of Chongqing Municipality, Chongqing University of Science and Technology, Chongqing 401331, China
2 China National Petroleum Corporation Great Wall Drilling Company, Beijing 100101, China
3 The Key Laboratory of Enhanced Oil and Gas Recovery of Education Ministry, Northeast Petroleum University, Daqing 163000, China
* Correspondence: yangshenyaocq@163.com (S.Y.); 2012020@cqust.edu.cn (W.Y.)

Abstract: Supercritical multicomponent thermal fluid (scMCTF) is a novel medium with great potential for heavy oil thermal recovery. The production rate of scMCTF will affect the injection efficiency of thermal fluid, and then affect the development effect of thermal recovery. However, at present, there are few reports on the production rate of each component of scMCTF, and their understanding is not clear. According to the existing production rate data of supercritical water (scH_2O) gasification products, based on the generation mechanism of scMCTF, the production rate of thermal fluid generation products under different generation conditions was calculated, and its influencing factors were identified. The results show the following: (1) The factors affecting the production rate of scMCTF generation products can be divided into three categories: reaction raw material factors, reaction condition factors, and catalytic factors. (2) The hydrocarbon number of raw material, reaction temperature, reaction time, and catalyst concentration were positively correlated with the production rate of the product. (3) The concentration of the reaction raw material is negatively correlated with the production rate of the product. The higher the concentration of the raw material is, the lower the concentration of H_2O is, and the steam reforming reaction is inhibited, which leads to the decrease in the production rate. (4) The effect of reaction pressure and catalyst load on the product is not significant. (5) The reaction product production rate increased first and then decreased with the ratio of H_2O to oil in the raw material emulsion and the ratio of preheated H_2O to raw material discharge. (6) The effect of metal salt catalysts is relatively stable, and the catalytic effect of simple metal catalysts is significantly different under the action of different types of accelerators, so it is necessary to study the degree of synergization of different accelerators on the catalytic effect. The results can lay a foundation for the subsequent experimental and theoretical research design.

Keywords: supercritical multicomponent thermal fluid (scMCTF); yield; productivity; factors; heavy oil thermal recovery

1. Introduction

The development of unconventional oil and gas such as heavy oil, shale gas, and oil require technological innovation [1,2]. scMCTF injection is a novel thermal recovery technology for heavy oil proposed in recent years [3,4]. scMCTF is composed of supercritical water (scH_2O), supercritical nitrogen (scN_2), and supercritical carbon dioxide ($scCO_2$), and it possesses high solubility, high diffusivity, and high reactivity [5]. Compared to traditional steam mediums, it can significantly enhance heavy oil recovery [6,7].

scMCTF is produced by gasification reaction and oxidation reaction of organic matter under high temperature and high pressure. The production rate of scMCTF is influenced by various reaction conditions such as feedstock, temperature, and pressure [8–10], and affects the thermal fluids' injection rate and injection volume. The injection rate and volume are key parameters in the thermal fluids injection process for heavy oil recovery, and different injection rate and volume will bring disparate oil recovery. Precisely controlling these parameters according to the pre-set plan is fundamental to ensuring the effectiveness of thermal recovery [11,12]. Therefore, accurately understanding the influence of different reaction conditions on the production rate of scMCTF is crucial for the efficient development of heavy oil resources using this technology.

However, there is currently a lack of research on the production rate of scMCTF, leading to insufficient understanding in this area. In light of this, based on existing research data on the production rate of hydrocarbon organics commonly used in oil fields under scH_2O conditions, and grounded in the theory of scMCTF generation, this study calculates the production rate of scMCTF under different reaction conditions, analyzes the variation patterns of production rate with different factors, and identifies the key reaction conditions influencing the production rate. The research findings can provide a reference for future research approaches and the design of methodologies for studying the production rate of scMCTF.

2. scMCTF Yield Calculation

The typical generation process of scMCTF includes two stages: organic matter gasification and oxidation of gasification products [13]. The reaction equations for the gasification process in the first stage (Equations (1)–(3)) and the oxidation process in the second stage (Equations (4)–(6)) are as follows. Through a literature search, 367 data samples of gasification production rates under scH_2O conditions for commonly used hydrocarbon organics in oil fields under different reaction conditions were obtained (Table 1). These samples show the production rate of gasification products in a scH_2O environment, which corresponds to the production rate of the first-stage products in the generation process of scMCTF.

$$C + H_2O \rightarrow CO + H_2 \tag{1}$$

$$CO + H_2O \rightarrow CO_2 + H_2 \tag{2}$$

$$CO + 3H_2 \rightarrow CH_4 + H_2O \tag{3}$$

$$2H_2 + O_2 \rightarrow 2H_2O \tag{4}$$

$$2CO + O_2 \rightarrow 2CO_2 \tag{5}$$

$$CH_4 + 2O_2 \rightarrow CO_2 + 2H_2O \tag{6}$$

Table 1. Sample data on the production rate of gasification products from scH_2O treatment of commonly used hydrocarbon organics in oilfields.

Researcher	Time	Reference	Reaction Raw Material	Raw Material Source	Sample Size, Group
Susanti et al.	2014	[14]	iso-octane, n-octane, n-decane, n-dodecane, n-hexadecane	Daejung Chemistry, Korea, Japan Yakuri Pure Chemical, Japan Kanto Chemical, UK Alfa Aesar Company	9
Wang et al.	2015	[15]	Oily sludge	An oil field production site in China	22
Kou et al.	2018	[16]	Diesel oil	China Petroleum and Chemical Corporation	21
Xu et al.	2019	[17]	Diesel oil	China Petroleum and Chemical Corporation	12
Zhang et al.	2019	[18]	n-undecane, n-hexadecane	Shanghai Macklin	10
Duan	2020	[19]	Oily sludge	An oil field production site in China	20
Peng et al.	2021	[20]	Oily sludge	Bohai Oilfield	11

Table 1. *Cont.*

Researcher	Time	Reference	Reaction Raw Material	Raw Material Source	Sample Size, Group
Liu	2021	[21]	Oily sludge	Changqing Oilfield	41
Kou et al.	2021	[22]	Diesel oil	China Petroleum and Chemical Corporation	2
Lei	2022	[23]	Oily sludge	Changqing Oilfield	36
Peng et al.	2022	[24]	Oily sewage	Bohai Oilfield	22
Kou et al.	2022	[25]	Diesel oil	China Petroleum and Chemical Corporation	49
Xu et al.	2022	[26]	Diesel oil	China Petroleum and Chemical Corporation	28
Wang et al.	2023	[27]	Oily sludge	Changqing Oilfield	23
Peng et al.	2023	[28]	Oily sewage	China National Offshore Oil Corporation	17
Li et al.	2023	[29]	Oily sludge	Karamay Oilfield	8
Peng et al.	2023	[30]	Oily sewage	Bohai Oilfield	21
Li et al.	2023	[31]	Oily sludge	An oil field in Xinjiang, China	15

Based on the production rate data of the first-stage products and using the theory of scMCTF generation along with the principle of mass conservation, the production rate of each component in the scMCTF can be calculated. The calculation is based on the following assumptions: ① The organic feedstock and H_2O introduced into the gasification reactor react completely. ② The reverse reactions of the water-gas shift reaction and the methanation reaction can be neglected [32,33]. ③ The oxygen (O_2) injected into the oxidation reactor reacts completely with the gasification products from the first stage. ④ The small amounts of C_2H_n gases in the products are ignored. ⑤ The mass fraction of O_2 in the air is assumed to be 76.7%, and nitrogen (N_2) is 23.3%. The calculation process is shown in Figure 1.

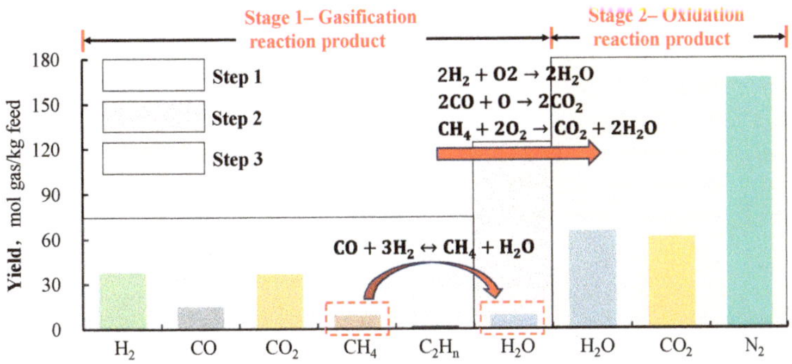

Figure 1. Schematic diagram for production rate of calculation in the scMCTF (data sourced from [15]).

Step 1: Obtain and organize the production rate data of the first-stage products from previous literature.

Step 2: Since previous studies cooled the products before analysis, the reaction-generated H_2O was not included in the product analysis. However, in the methanation reaction, the molar ratio of methane to H_2O in the products is 1:1, which is used to determine the production rate of H_2O in the first-stage gasification reaction.

Step 3: Based on the production rate of the first-stage products and using the oxidation reaction equations from the second stage, the production rate of scH_2O and $scCO_2$ in the second-stage reaction products can be determined. The N_2 production rate is then determined based on the O_2 consumption in the second-stage reaction and the O_2–N_2 ratio in the air.

By following these steps, the production rate of the scMCTF can be established.

3. Influencing Factors of scMCTF Yield

The factors influencing the production rate of scMCTF components can be categorized into three types: ① Feedstock factors, including feedstock type, feedstock concentration, and water-to-oil ratio in the feedstock emulsion. ② Reaction condition factors, including reaction pressure, reaction temperature, reaction time, and flow rate ratio of preheated H_2O to feedstock during the reaction process. ③ Catalytic factors, including catalyst type, concentration, and loading.

3.1. Reaction Material Factor

As can be seen from Figure 2, each component's yield of scMCTF generally increases with the number of carbon atoms in the reaction raw material. Hydrocarbons are composed of carbon and hydrogen atoms, and the greater the number of carbon and hydrogen atoms, the larger the molecular weight of the feedstock, resulting in higher scH_2O and $scCO_2$ content in the products after the complete reaction. Additionally, since feedstocks with larger molecular weights generate more hydrogen and methane during the gasification stage, more O_2 is consumed during the oxidation stage, leading to an increase in the amount of N_2 remaining in the air.

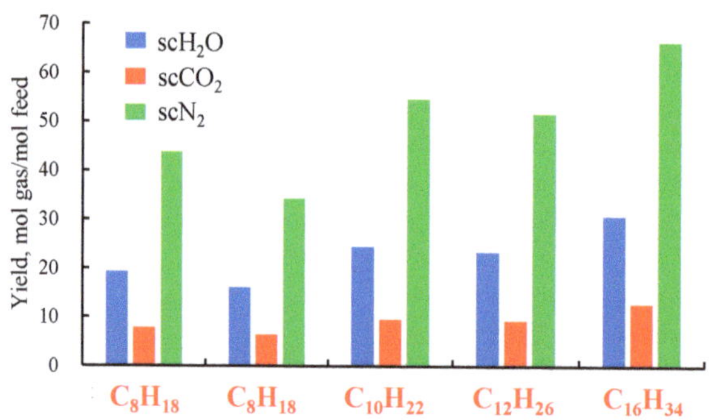

Figure 2. Production rates of scMCTF from different feedstocks (data from [14]).

Figure 3 shows the negative impact of feedstock concentration on the production rate of each component in the scMCTF. The higher the feedstock concentration, the lower the production rate. This is because the steam reforming reaction is sensitive to H_2O concentration [34] and a decrease in H_2O concentration inhibits the steam reforming reaction, leading to incomplete reactions when the feedstock concentration is high [32]. As observed from the overall data distribution in Figure 3, within the low concentration range, the production rate of components decreases as the concentration increases. As the concentration continues to rise, the production rate gradually stabilizes, suggesting the existence of a threshold concentration. Once the feedstock concentration exceeds this threshold, the impact of concentration on production rate becomes less significant. Therefore, an appropriate feedstock concentration is crucial: if the concentration is too low, the amount of each component in the scMCTF may be insufficient, while an excessively high concentration can inhibit the reaction.

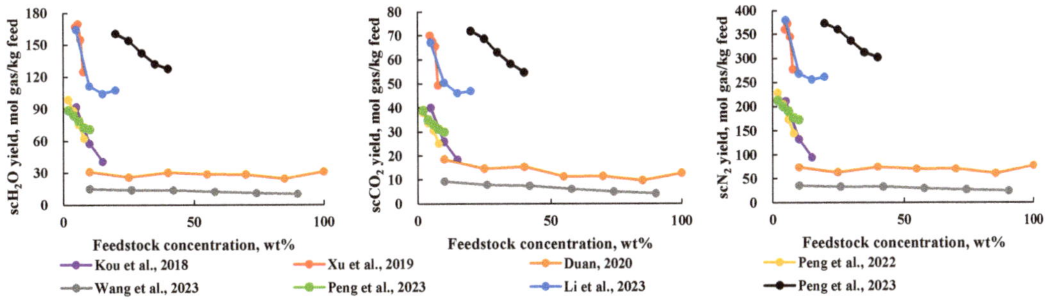

Figure 3. Relationship between reaction feedstock concentration and production rate of various components [16,17,19,24,27–30].

Catalysts are widely used in reactions to enhance production rate [16,17,21,22,25]. Commonly used metal salt catalysts and elemental metal catalysts are generally difficult to dissolve in hydrocarbons. Therefore, the feedstock, H_2O, and catalysts are often prepared into an emulsion to improve catalyst dispersion and thus increase catalytic efficiency. Xu [17], using a continuous reaction system (where the feedstock and products are continuously input and output), studied the effect of the water-to-oil ratio in the emulsion on the component production rate. They found that the component production rate first increased and then decreased as the water-to-oil ratio of the emulsion increased (Figure 4). They attributed this to the dual effects of reaction temperature changes and the micro-explosion phenomenon. On one hand, a higher water-to-oil ratio means a higher H_2O content in the emulsion, which can lower the local temperature in the reaction zone, leading to an increase in side reactions and negatively affecting the reaction outcome. On the other hand, since H_2O has a lower boiling point compared to hydrocarbon feedstocks, it reaches a superheated state more quickly when heated in the reactor. This causes micro-explosions, breaking the feedstock into smaller droplets, thereby enhancing the reaction efficiency. Therefore, optimizing the water-to-oil ratio in the emulsion can improve the component production rate. Additionally, since the water-to-oil ratio in the emulsion affects the feedstock concentration, these two factors should be considered together in the optimization process.

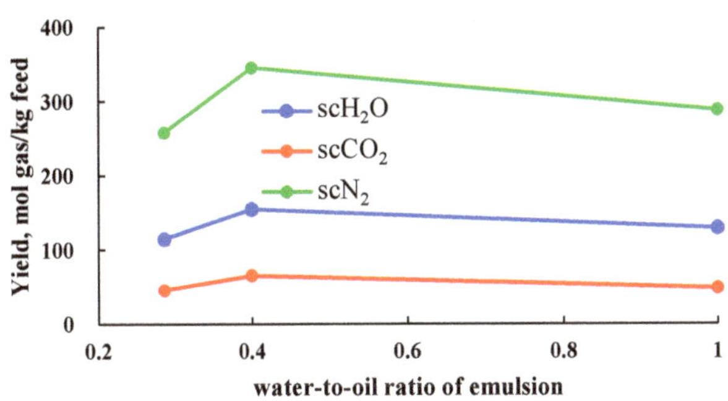

Figure 4. Relationship between water-to-oil ratio and production rate of various components (data from [17]).

3.2. Reaction Conditioning Factor

Researchers believe that reaction pressure has an insignificant impact on product production rate, which has led to the adoption of a fixed reaction pressure in studies [14–18,23–26]. Some researchers have verified this view by studying the effect of reaction pressure on production rate [19,27,28]. As shown in Figure 5, the production rate of products shows almost no change with increasing pressure, with only a slight effect on $scCO_2$ production rate [19]. This effect has not been deeply analyzed, but considering the findings and viewpoints of most researchers, it is likely due to experimental error.

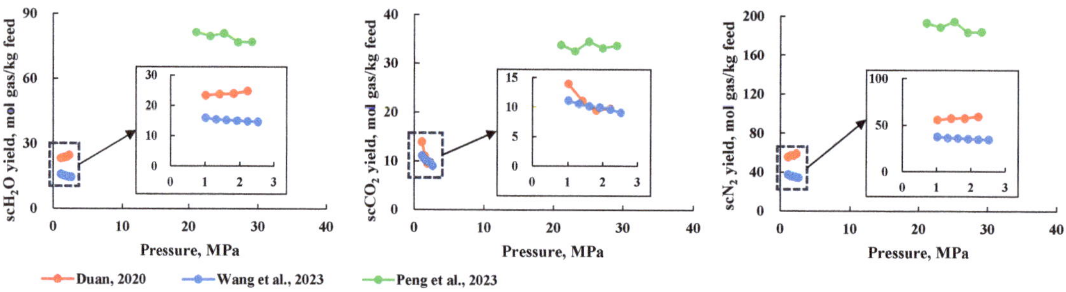

Figure 5. Relationship between pressure and production rate of various components [19,27,28].

Reaction temperature is one of the most important factors affecting product production rate. During the generation of scMCTF, the production rate of all components increases with rising temperature (Figure 6), and the trend of production rate growth with temperature is generally consistent across components. Higher temperatures favor the reaction process and product formation. However, higher temperatures also mean increased temperature requirements for the supercritical thermal fluid equipment and higher heating fuel costs. Therefore, it is necessary to lower the reaction temperature as much as possible while still ensuring adequate product production rate, to reduce equipment demands and fuel costs.

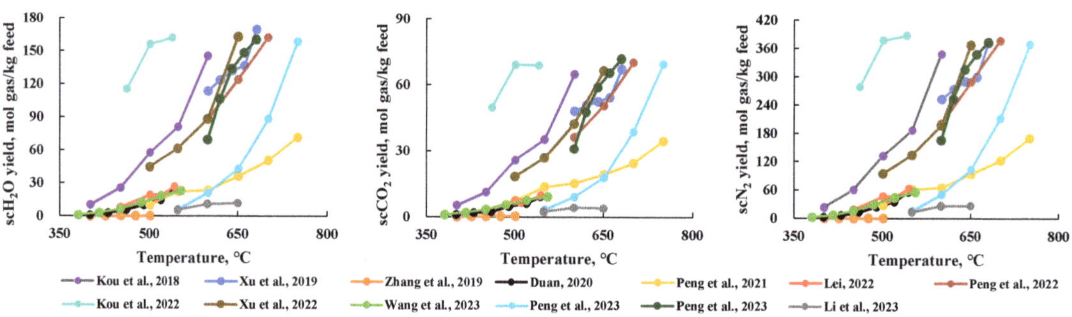

Figure 6. Relationship between temperature and production rate of various components [16–20,23–28,30,31].

In a batch reaction system (where feedstock input, reaction, and product output occur in intermittent cycles), the reaction time is usually sufficient to ensure a complete reaction of the feedstock. However, in a continuous reaction system, the time during which the feedstock comes into contact with scH_2O determines the extent of the reaction and the production rate of the products. As shown in Figure 7, the product production rate increases with a longer reaction time, but after sufficient reaction time, the rate of production rate increase slows down and eventually levels off. Generally, a reaction time of over 30 min is sufficient to ensure a complete reaction of the products. However, in continuous systems, where feedstock is continuously introduced and scMCTF is produced, reaction time for the feedstock is typically shorter [14,26]. Therefore, to achieve a complete reaction within a shorter time, it is necessary to increase the reaction temperature or add catalysts to reduce the time required for the feedstock to fully react.

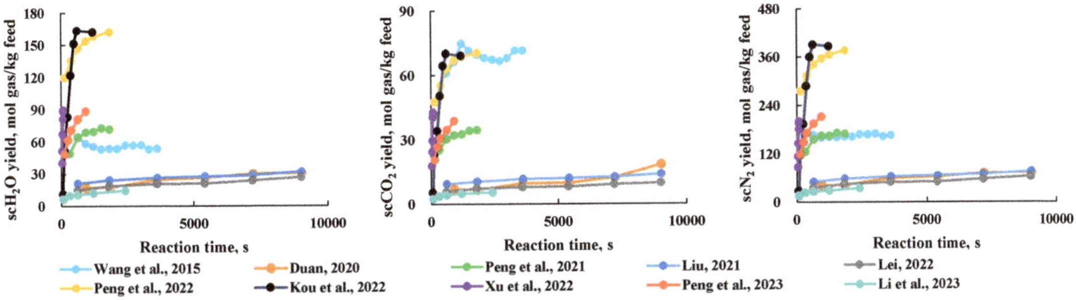

Figure 7. Relationship between reaction time and production rate of various components [15,19–21,23–26,28,31].

In a continuous reaction system, preheated H_2O in a supercritical state and the feedstock are injected into the reactor through different nozzles at a specific ratio to react. Therefore, the "preheated water-to-feedstock flow rate ratio" factor applies only to continuous reaction systems and is not relevant to batch systems. As shown in Figure 8, as the preheated water-to-feedstock flow rate ratio increases, the product production rate first increases and then decreases, indicating the existence of an "optimal flow rate ratio" during this process. This pattern is primarily due to two reasons. First, a larger flow of preheated H_2O enhances heat and mass transfer within the reactor, allowing the feedstock to quickly absorb heat and react under high flow conditions. However, an excessively high flow rate shortens the residence time of the reactants in the reactor, which may cause the feedstock to exit the reactor before fully reacting. Additionally, Figure 8 shows that the "optimal flow rate ratio" varies across different studies, which is due to differences in experimental conditions. Therefore, determining the specific "optimal flow rate ratio" under given conditions is crucial for optimizing the production rate.

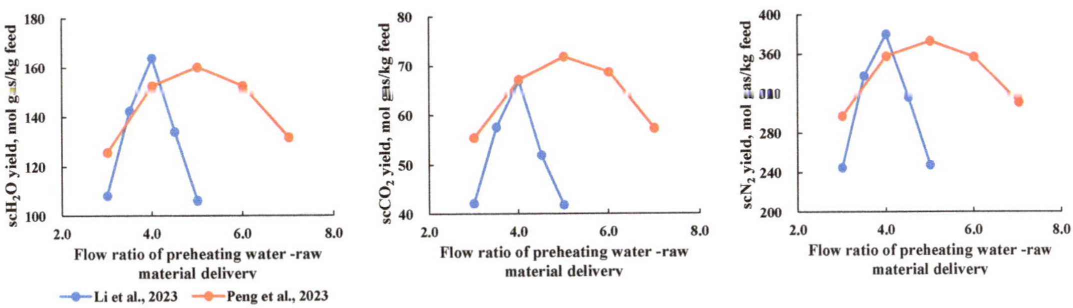

Figure 8. Relationship between flow ratio of preheating water and raw material delivery, and components production [29,30].

3.3. Catalytic Factor

Metal salt catalysts and elemental metal catalysts are the two most commonly used types of catalysts in scH_2O gasification reactions. Among them, elemental metal catalysts can be combined with various promoters to enhance their activity. As shown in Figure 9, all types of catalysts effectively increase production rates. Metal salt catalysts exhibit a significant catalytic effect, with relatively small differences in performance between different types of metal salts [21,30], making their catalytic effects quite stable. On the other hand, the catalytic performance of elemental metal catalysts varies greatly depending on the type of promoter used. For example, in the study by [16], the catalytic effect of a Ni catalyst supported on ZrO_2 showed significant variation when different promoters were added. Compared to the catalyst without a promoter, the catalyst with Co as a promoter nearly doubled the product production rate, while using Y as a promoter resulted in a catalytic effect that was even worse than that without any promoter.

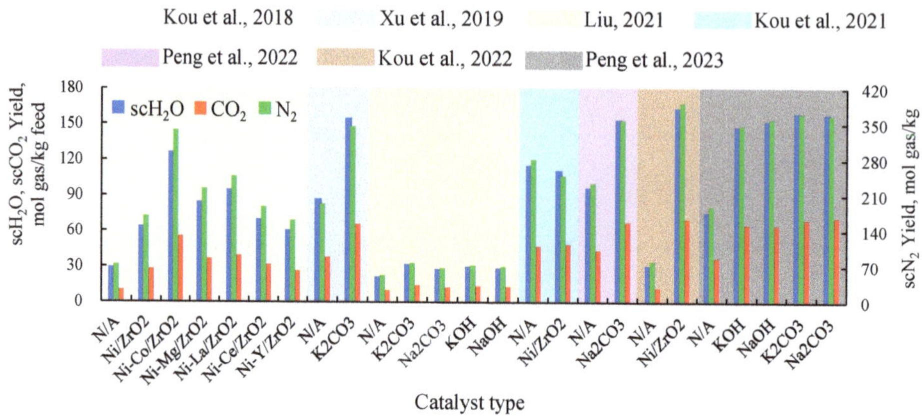

Figure 9. Production rate of various components under different catalysts [16,17,21,22,24,25,30].

Catalysts can increase product production rate by promoting steam reforming reactions and water-gas shift reactions. The higher the catalyst mass concentration, the greater the amount of active catalytic material, which in turn enhances the reaction. As shown in Figure 10, the production rate of each component in the scMCTF increases with the rise in catalyst mass concentration. However, the extent of the production rate increase varies among different studies. For instance, in study [21], raising the catalyst mass concentration from 1 wt% to 3 wt% led to increases in component production rates of 15.7%, 20.5%, and 15.7%, respectively. In another study [23], increasing the catalyst mass concentration from 1 wt% to 5 wt% resulted in production rate increases of approximately 7.0% for scH_2O, 8.2% for $scCO_2$, and 6.5% for scN_2. In a third study [30], increasing the catalyst mass concentration from 4 wt% to 8 wt% resulted in production rate increases of 30.3% for scH_2O, 28.9% for $scCO_2$, and 28.1% for scN_2.

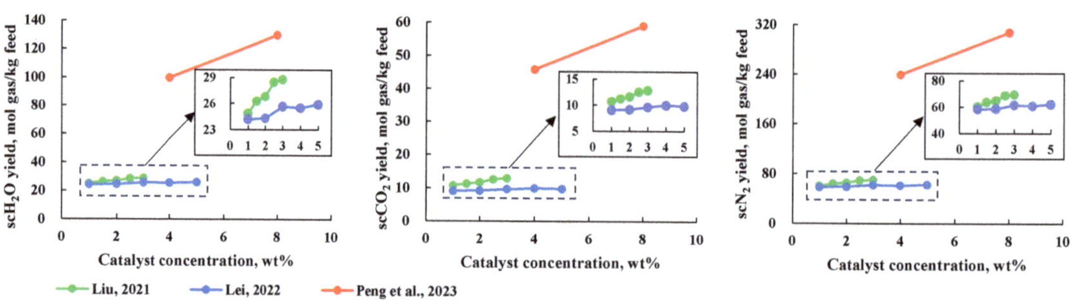

Figure 10. Relationship between catalyst mass concentration and production tare of various components [21,23,30].

Some elemental metal catalysts tend to deactivate at high temperatures [35]. To retain the catalytic activity of elemental metals and mitigate deactivation under high-temperature conditions, combining elemental metals with metal oxide supports to create supported catalysts is effective. The catalyst loading refers to the mass fraction of the active phase in the supported catalyst. Lei [23] prepared a supported catalyst with Ni as the active phase and Al_2O_3 as the support, studying the impact of catalyst loading on production rate. The study revealed that catalyst loading has a minimal effect on product rate, which is consistent with Lei's findings on catalyst concentration (Figure 11). This may be due to a phenomenon similar to the catalyst concentration threshold, where catalyst loading also has a threshold. In the study [23], the loading likely exceeded this threshold, resulting in an insignificant change in production rate with varying loading levels.

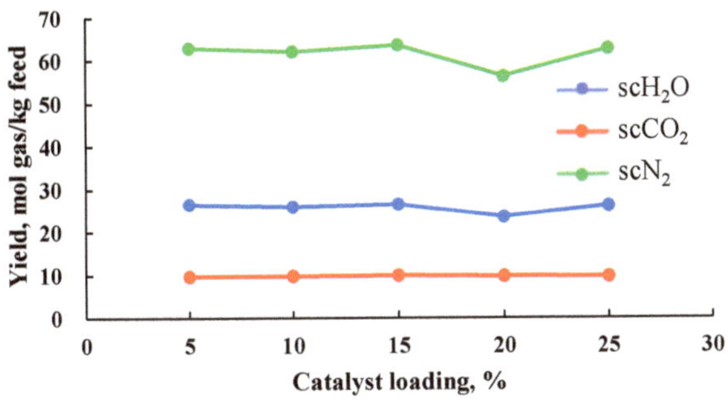

Figure 11. Relationship between the amount of catalyst load and components production rate (data from [23]).

4. Discussion

We mentioned above that the previous work focused on the yield of organic gasification products, but did not pay attention to the yield of scMCTF [14–31], because the concept of scMCTF has been put forward in the past five years [3–7], and there are very few studies on it. Meanwhile, there is no relevant report on the yield of scMCTF so far. Therefore, it is a novel work which provides a first view of the factors affecting the scMCTF yield in generation processes. This view can help scholars understand the factors that affect the yield of scMCTF and how the yield changes with them, which is the basis for conducting further experimental or simulation research.

The reference provided by this work is also limited because it has some irremediable defects. The thermal fluid yield obtained in this work is calculated theoretically based on the yield of organic matter gasification products in previous works, which means that research conditions are not uniform. In other words, the data for each of the researchers we cite may have been obtained under very different experimental conditions [14–31]. For example, in Section 3.2, the temperature is shown as an important factor (Figure 6). If the effect of temperature on yield is to be studied, factors other than temperature such as pressure, catalyst, etc., should be consistent. However, we cannot guarantee that other factors are the same, which leads to the possibility that other factors may have an additional effect on the temperature effect. Therefore, this work can only provide a qualitative rule, but cannot give specific and quantitative suggestions.

5. Conclusions and Prospects

The reaction conditions during the formation of scMCTF significantly affect product production rate. However, there is currently a lack of research specifically focused on the production rate of components in scMCTF, leading to an incomplete understanding of the subject. This paper, drawing on similar studies of scH_2O gasification product production rates and based on the theory of scMCTF formation, calculates the product production rate under different reaction conditions and investigates the factors influencing the production rate of each component. The following conclusions were drawn:

(1) The factors influencing the production rate of scMCTF components can be classified into three categories: ① feedstock factors, including feedstock type, feedstock concentration, and the water-to-oil rate of the feedstock emulsion; ② reaction conditions, including reaction pressure, reaction temperature, reaction time, and preheated water-

to-feedstock flow rate ratio during the reaction process; ③ catalytic factors, including catalyst type, catalyst concentration, and catalyst loading.

(2) There is a positive correlation between product production rate and the number of carbon-hydrogen atoms in the feedstock, reaction temperature, reaction time, and catalyst concentration. When the reaction is complete, the higher the number of carbon-hydrogen atoms in the feedstock, the higher the production rate; higher reaction temperatures lead to higher product production rates; as reaction time extends, the product production rate initially increases and then stabilizes; and higher catalyst concentration results in higher product production rates.

(3) There is a negative correlation between feedstock concentration and product production rate. Higher feedstock concentrations result in lower H_2O concentrations, which inhibit the steam reforming reaction and subsequently reduce the production rate.

(4) Reaction pressure and catalyst loading have a minimal impact on product production rate, with production rates remaining largely unchanged as pressure and catalyst loading vary.

(5) Product production rate exhibits an initial increase followed by a decrease with changes in both the water-to-oil ratio of the feedstock emulsion and the preheated water-to-feedstock flow rate ratio. The change in product production rate with the water-to-oil ratio is driven by both temperature variation and micro-explosion effects, while the change in production rate with the preheated water-to-feedstock flow rate ratio is due to differences in local reactor temperatures and the residence time of reactants.

(6) Metal salt catalysts and elemental metal catalysts are the two most commonly used catalysts in scH_2O gasification reactions. Metal salt catalysts generally exhibit stable performance. The catalytic effectiveness of elemental metal catalysts, however, varies significantly depending on the type of promoter used. Therefore, it is essential to study the enhancement effects of different promoters on catalytic activity.

Author Contributions: Methodology, J.T. and M.D.; Formal analysis, F.X. and D.K.; Investigation, S.Y.; Writing – original draft, W.Z.; Funding acquisition, Z.Q. and W.Y. All authors have read and agreed to the published version of the manuscript.

Funding: This work is supported by National Natural Science Foundation of China (Grant number 52004048, U22B2074, 52104025), the Natural Science Foundation of Chongqing Municipality, China (Grant number cstc2020jcyj msxmX0856), Science and Technology Research Program of Chongqing Municipal Education Commission (KJQN202001508), Natural Science Foundation of Chongqing, China (CSTB2022NSCQ-MSX0858), the Science and Technology Research Program of Chongqing Municipal Education Commission (Grant No. KJZD-M202301501), Talent Plan project of Chongqing Municipal (cstc2022ycjh-bgzxm0055).

Data Availability Statement: The data presented in this study are available on request from the corresponding author. The data are not publicly available due to privacy restrictions.

Conflicts of Interest: Authors Xu, F. and Kong, D. were employed by the company China National Petroleum Corporation. The remaining authors declare that the research was conducted in the absence of any commercial or financial relationships that could be construed as a potential conflict of interest.

References

1. Li, Q.C.; Li, Y.D.; Cheng, Y.F.; Li, Q.; Wang, F.; Wei, J.; Liu, Y.; Zhang, C.; Song, B.; Yan, C.; et al. Numerical simulation of fracture reorientation during hydraulic fracturing in perforated horizontal well in shale reservoirs. *Energy Sources Part A Recovery Util. Environ. Eff.* **2018**, *40*, 1807–1813. [CrossRef]
2. Li, Q.C.; Cheng, Y.F.; Li, Q.; Zhang, C. Establishment and evaluation of strength criterion for clayey silt hydrate-bearing sediments. *Energy Sources Part A Recovery Util. Environ. Eff.* **2018**, *40*, 742–750. [CrossRef]
3. College of Petroleum and Natural Gas Engineering, Chongqing University of Science and Technology. The Key Project Launch and Program Review Meeting of the National Self-Science Fund Enterprise Innovation and Development Joint Fund of Our School Was Held [EB/OL]. 2023. Available online: https://sgy.cqust.edu.cn/info/1221/2142.htm (accessed on 6 December 2023).

4. State Key Laboratory of Dynamic Engineering Multiphase Flow, Xi'an Jiaotong University. Academician Guo Liejin, Director of the Laboratory, and Young Teacher Zhao Qiuyang, Have Made Important Progress in the Research on the Mechanism of Supercritical Fluid Development of Unconventional Oil, and Their Results Have Been Selected as a Journal Cover Article [EB/OL]. 2023. Available online: http://mfpe.xjtu.edu.cn/info/1077/6897.htm (accessed on 6 December 2023).
5. Zhao, Q.Y.; Jin, H.; Xu, J.L.; Peng, Z.Y.; Wang, H.; Guo, L. Theory and Technology of Multi-Component Supercritical Thermal Fluid Generation and In-Situ Conversion Technique for Heavy Crude Oil Resources Exploitation. *J. Xi'an Jiaotong Univ.* **2023**, *57*, 31–45.
6. Li, X.Y.; Sun, X.F.; Cai, J.M.; Zhang, Q.; Pan, X.; Zhang, Y. Experimental investigation on supercritical multi-thermal fluid flooding using a novel 2-dimensional model. *Energy* **2023**, *283*, 129136. [CrossRef]
7. Huang, Z.J.; Zhao, Q.Y.; Chen, L.; Miao, Y.; Wang, H.; Jin, H.; Guo, L. Fundamentals of Enhanced Heavy Oil Recovery by Supercritical Multi-Component Thermal Fluid Flooding. *J. Eng. Thermophys.* **2022**, *43*, 974–981.
8. Yan, Z.Y.; Tan, X.Y. Hydrogen generation from oily wastewater via supercritical water gasification (SCWG). *J. Ind. Eng. Chem.* **2015**, *23*, 44–49.
9. Sanchez-Hernandez, A.M.; Martin-Sanchez, N.; Sanchez-Montero, M.J.; Izquierdo, C.; Salvador, F. Effect of pressure on the gasification of dodecane with steam and supercritical water and consequences for H_2 production. *J. Mater. Chem. A* **2018**, *6*, 1671. [CrossRef]
10. Wang, G.Y.; Li, J.S.L.; Li, X.J.; Kou, J.; Ge, Z.; Li, L.; Peng, P.; Guo, L. Experimental study on supercritical water gasification of oily sludge using a continuous two-step method. *J. Hazard. Mater.* **2023**, *455*, 131619. [CrossRef]
11. Yang, G.L.; Li, Y.H.; He, H.Z. Research and Application of Multi-Component Thermal Fluid Stimulation in the Extra-Heavy Oil Reservoir. *Spec. Oil Gas Reserv.* **2020**, *27*, 103–107.
12. Zhong, L.G.; Jiang, Y.; Ma, S. Physical and numerical simulation of multi-component-thermal-fluid-assisted gravity drainage in deep and extra-heavy oil reservoirs offshore. *China Offshore Oil Gas* **2015**, *27*, 68–73.
13. Xu, J.; Peng, Z.; Rong, S.; Zhao, Q.; Jin, H.; Guo, L.; Zhou, T.; Zhang, X. Optimal retrofit of a novel multi-component supercritical thermal fluid generation system via thermodynamic analysis. *Appl. Therm. Eng.* **2023**, *219*, 119511. [CrossRef]
14. Susanti, R.F.; Dianningrum, L.W.; Yum, T.; Kim, Y.; Lee, Y.-W.; Kim, J. High-production rate hydrogen production by supercritical water gasification of various feedstocks: Alcohols, glucose, glycerol and long-chain alkanes. *Chem. Eng. Res. Des.* **2014**, *92*, 1834–1844. [CrossRef]
15. Wang, F.Q.; Zhu, S.J.; Gong, X.L. Gasification of Oily Sludge in Supercritical Water. *Oxid. Commun.* **2015**, *38*, 1391–1400.
16. Kou, J.J.; Xu, J.J.; Jin, H.; Guo, L.; Zhang, D.; Cao, W. Evaluation of modified Ni/ZrO_2 catalysts for hydrogen production by supercritical water gasification of oil-containing wastewater. *Int. J. Hydrogen Energy* **2018**, *43*, 13896–13903. [CrossRef]
17. Xu, J.; Kou, J.; Guo, L.; Jin, H.; Peng, Z.; Ren, C. Experimental study on oil-containing wastewater gasification in supercritical water in a continuous system. *Int. J. Hydrogen Energy* **2019**, *44*, 15871–15881. [CrossRef]
18. Zhang, H.; Wang, C.Y.; Cheng, C.; Zhong, J. Study on the Reaction of Saturated Hydrocarbons of Oily Sludge in Supercritical Water. *Environ. Sci. Technol.* **2019**, *32*, 20–23+29.
19. Duan, Y.W. Experimental Study on Hydrogen Production by Supercritical Water Gasification of Oily Sludge. Master's Thesis, Xi'an Shiyou University, Xi'an, China, 2020.
20. Peng, P.; Guo, S.; Li, L.; Jin, H.; Ge, Z.; Guo, L. Supercritical water gasification mechanism of polymer-containing oily sludge. *Int. J. Hydrogen Energy* **2021**, *46*, 26834–26847. [CrossRef]
21. Liu, X.B. Study on Mechanism of Hydrogen Production by Supercritical Water Gasification of Oily Sludge under Alkaline. Master's Thesis, Xi'an Shiyou University, Xi'an, China, 2021.
22. Kou, J.J.; Lei, Y.; Li, G.L.; Cheng, K.; Wang, R.; Zhang, D.; Jin, H.; Guo, L. Structural effect of ZrO_2 on supported Ni-based catalysts for supercritical water gasification of oil-containing wastewater. *Int. J. Hydrogen Energy* **2021**, *46*, 12874–12885. [CrossRef]
23. Lei, D.L. Study on Heterogeneous Gasification Law of Oily Sludge in Supercritical Water System. Master's Thesis, Xi'an Shiyou University, Xi'an, China, 2022.
24. Peng, Z.Y.; Rong, S.Q.; Xu, J.J.; Jin, H.; Zhang, J.; Shang, F.; Guo, L. Reaction pathways and kinetics for hydrogen production by oilfield wastewater gasification in supercritical water. *Fuel* **2022**, *314*, 123135. [CrossRef]
25. Kou, J.J.; Feng, H.F.; Wei, W.W.; Wang, G.; Sun, J.; Jin, H.; Guo, L. Study on the detailed reaction pathway and catalytic mechanism of a Ni/ZrO_2 catalyst for supercritical water gasification of diesel oil. *Fuel* **2022**, *312*, 122849. [CrossRef]
26. Xu, J.; Cheng, Z.; Ren, C.; Yi, L.; Wei, W.; Jin, H.; Guo, L. Supercritical water gasification of oil-containing wastewater with a homogeneous catalyst: Detailed reaction kinetic study. *Int. J. Hydrogen Energy* **2022**, *47*, 25541–25554. [CrossRef]
27. Wang, Y.L.; Wang, S.Z.; Qi, H.Y.; Jiang, H.; Duan, Y. Characteristics and prediction model of hydrogen production of oily sludge by supercritical water gasification. *Int. J. Hydrogen Energy* **2023**, *48*, 11191–11204. [CrossRef]
28. Peng, Z.Y.; Xu, J.J.; Rong, S.Q.; Zhang, M.; Wang, L.; Jin, H.; Guo, L. Clean treatment and resource utilization of oilfield wastewater using supercritical water gasification. *J. Clean. Prod.* **2023**, *411*, 137239. [CrossRef]
29. Li, L.H.; Wang, G.Y.; Li, X.J.; Wang, L.; Zhang, J.; Cheng, K.; Peng, P.; Cao, W.; Jin, H.; Guo, L. Experimental study on alkali catalytic gasification of oily sludge in supercritical water with a continuous reactor. *J. Environ. Manag.* **2023**, *327*, 116957. [CrossRef]
30. Peng, Z.Y.; Rong, S.Q.; Xu, J.J.; Luo, K.; Zhang, J.; Jin, H.; Guo, L. Hydrogen production from oilfield wastewater by gasification in supercritical water with a continuous system. *Fuel* **2023**, *344*, 128094. [CrossRef]

31. Li, L.H.; Li, X.J.; Cao, W. Reaction pathway and kinetics study on supercritical water gasification of oily sludge. *J. Anal. Appl. Pyrolysis* **2023**, *170*, 105920. [CrossRef]
32. Jin, H.; Lu, Y.; Liao, B.; Guo, L.; Zhang, X. Hydrogen production by coal gasification in supercritical water with a fluidized bed reactor. *Int. J. Hydrogen Energy* **2010**, *35*, 7151–7160. [CrossRef]
33. Lan, R.H.; Jin, H.; Guo, L.J.; Ge, Z.; Guo, S.; Zhang, X. Hydrogen Production by Catalytic Gasification of Coal in Supercritical Water. *Energy Fuels* **2014**, *28*, 6911–6917. [CrossRef]
34. Guo, S.M.; Guo, L.J.; Cao, C.Q.; Yin, J.; Lu, Y.; Zhang, X. Hydrogen production from glycerol by supercritical water gasification in a continuous flow tubular reactor. *Int. J. Hydrogen Energy* **2012**, *37*, 5559–5568. [CrossRef]
35. Nguyen, H.T.; Lu, H.; Kobayashi, E.; Ishikawa, T.; Komiyama, M. Raney-Nickel Catalyst Deactivation in Supercritical Water Gasification of Ethanol Fermentation Stillage and its Mitigation. *Top. Catal.* **2014**, *57*, 1078–1084. [CrossRef]

Disclaimer/Publisher's Note: The statements, opinions and data contained in all publications are solely those of the individual author(s) and contributor(s) and not of MDPI and/or the editor(s). MDPI and/or the editor(s) disclaim responsibility for any injury to people or property resulting from any ideas, methods, instructions or products referred to in the content.

energies

MDPI

Article

Gas Production Prediction Model of Volcanic Reservoir Based on Data-Driven Method

Haijie Zhang [1], Junwei Pu [1], Li Zhang [1], Hengjian Deng [1], Jihao Yu [2], Yingming Xie [3], Xiaochang Tong [4], Xiangjie Man [4] and Zhonghua Liu [2,4,*]

1 Chong Qing Shale Gas Exploration and Development, Co., Ltd., Chongqing 401121, China; zhang_hj@petrochina.com.cn (H.Z.); pujunwei@petrochina.com.cn (J.P.); li_zhang@petrochina.com.cn (L.Z.); dlts_denghj@petrochina.com.cn (H.D.)
2 School of Petroleum and Natural Gas Engineering, Chongqing University of Science and Technology, Chongqing 401331, China; 2022201054@cqust.edu.cn
3 CNOOC EnerTech-Drilling & Production, Co., Beijing 100028, China; xieyingming56@163.com
4 Pancasia Holding Co., Ltd., Chongqing 400000, China; cqlnytxc@163.com (X.T.); cqlnymxj@163.com (X.M.)
* Correspondence: liuzhonghua@cqust.edu.cn

Abstract: Based on on-site construction experience, considering the time-varying characteristics of gas well quantity, production time, effective reservoir thickness, controlled reserves, reserve abundance, formation pressure, and the energy storage coefficient, a data-driven method was used to establish a natural gas production prediction model based on differential simulation theory. The calculation results showed that the average error between the actual production and predicted production was 12.49%, and the model determination coefficient was 0.99, indicating that the model can effectively predict natural gas production. Additionally, we observed that the influence of factors such as reserve abundance, the number of wells in operation, controlled reserves, the previous year's gas production, formation pressure, the energy storage coefficient, effective matrix thickness, and annual production time on the annual gas production increases progressively as the F-values decrease. These insights are pivotal to a more profound understanding of gas production dynamics in volcanic reservoirs and are instrumental in optimizing stimulation treatments and enhancing resource recovery in such reservoirs and other unconventional hydrocarbon formations.

Keywords: production prediction model; volcanic reservoir; data-driven method; data nondimensionalization; dimension recovery

1. Introduction

Hydrocarbon production from unconventional reservoirs requires the integration of different technologies, including long lateral horizontal drilling and multi-stage, multi-cluster hydraulic fracture systems that activate natural fracture networks in unconventional formations [1,2]. Dark box approaches, which involve using data analytics techniques, have recently gained significant attention in many areas [3–5]. A variety of data analytics methods, such as machine learning [6], linear regression [7], and neural networks [8], are used for predicting gas and oil production.

Many scholars [9–13] from all over the world have applied linear regression methods in predicting gas and oil production. Zhou et al. [14] conducted a multiple regression model for 173 wells in the Marcellus Formation, predicting the one-year cumulative gas production as a function of the proppant mass, fracture fluid volume, number of stages, treatment rate, vertical depth, and lateral length. Grujic et al. [15] developed a predictive model for 172 wells in North America, forecasting oil, gas, and water production as a function of the volume, petrophysics, temperature, pressure, and geographical and completion parameters. Zhong et al. [16] predicted oil production in 476 wells in the Wolfcamp Formation as a function of the well and completion designs, using a multiple regression method and comparing it with other methods. Lolon et al. [17] predicted the cumulative oil production as a function of the stage of cementing, percentage of ceramic proppant, stage spacing, proppant intensity, water cut, fracturing fluid, and maximum treatment rate, by comparing some machine learning methods. Khanal et al. [18] forecasted the gas rate, cumulative gas, and condensate-to-gas ratio (CGR) using linear regression analysis for 335 simulations and 46 wells in the Eagle Ford Formation. Xue et al. [19] predicted the dynamic shale gas production rate as a function of hydraulic fracturing and geological properties using a multi-objective random forest regression method. Their sensitivity analysis revealed that the most influential parameters were the geological properties, including initial pressure and formation thickness. Johan et al. [20] employed a genetic algorithm to optimize completion designs, using several predictor variables, including the depth, lateral length, azimuth, total fluid, fluid intensity, total proppant, proppant intensity, and additional engineering features to account for the influence of neighboring wells and depletion.

The overall goal of this study is to develop a time variation characteristic prediction model that can accurately predict gas production in volcanic gas reservoirs. Specifically, this study aims to address the limitations of existing prediction models in processing the dynamic production data of unconventional gas reservoirs, which typically require a large amount of geological and permeability data. Traditional prediction methods require extensive numerical simulations. To overcome these challenges, we derived a new prediction model based on differential simulation theory and fitted the model using dimensionless gas production data obtained from volcanic gas reservoirs. In addition, we also calculated the error, coefficient of determination (R^2), and F-value between actual and predicted gas production to verify the effectiveness and accuracy of the model. Through this study, we hope to provide a new and more effective tool for predicting gas production in unconventional gas reservoirs, thereby providing a scientific basis for a development strategy and production parameter adjustments in oil and gas fields.

2. Methodology

2.1. Workflow

Data-driven prediction, a method grounded in using existing data for analyzing and modeling to forecast future trends, outcomes, or events, is utilized in this paper to develop a production forecast model with time-varying characteristics. The methodology is outlined in a flow chart, depicted in Figure 1, and involves four sequential steps: initial data preprocessing, which entails the collection and normalization of gas field data to neutralize dimensional influences, followed by two stages of accumulation to reduce historical data randomness and prepare the data for modeling; the subsequent establishment of a multiple linear regression model based on a differential equation; parameter estimation using the least squares method to ensure predictive accuracy; and finally, data dimension recovery following two reduction processes to facilitate the computation of projected gas production. This comprehensive approach ensures the creation of a precise and reliable prediction model, essential in the accurate forecasting of gas field production.

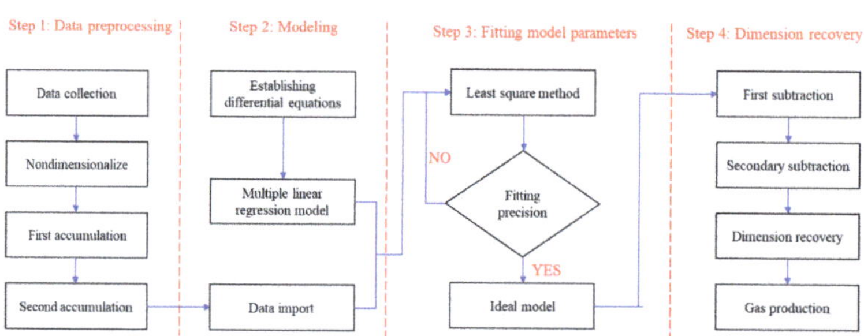

Figure 1. Flow chart of data-driven prediction.

2.2. *Data Preprocessing*

2.2.1. Data Collection

The indices for gas production data in volcanic gas reservoirs are bifurcated into two primary categories: the prediction index and the main control factor index. The prediction index is $Q(t)$ (gas production), and the main control factor index mainly contains U_1(t) (number of wells in operation), $U_2(t)$ (production time), $U_3(t)$ (effective thickness of reservoir), $U_4(t)$ (control reserves), $U_5(t)$ (reserve abundance), $U_6(t)$ (formation pressure), and $U_7(t)$ (energy storage factor). Then, all the data indexes are collected in an Excel sheet.

2.2.2. Data Nondimensionalization

To standardize the data and eliminate the discrepancies arising from different physical units among various parameters, each data index should be normalized using the following equation:

$$x^{(0)} = \frac{x}{\sum_1^n x} \tag{1}$$

where $x^{(0)}$ is the dimensionless value; x is the raw data of the gas field data; and n is the number of data.

2.2.3. Data Accumulation

To mitigate the randomness inherent to historical data and enhance the stability and reliability of the analysis, the dimensionless data are subjected to a first accumulation process. This is typically accomplished using the following equation:

$$x_i^{(1)} = \sum\nolimits_1^i x^{(0)} \tag{2}$$

where $x_i^{(1)}$ is the dimensionless value after the first accumulation; x is the value before normalization; and n is the number of data.

Then, the dimensionless data after the first accumulation can be accumulated for the second time to form a fitting sample library as the following equation:

$$x_i^{(2)} = \sum\nolimits_1^i x^{(1)} \tag{3}$$

2.3. *Modeling*

Differential simulation theory, a specialized data-driven approach, posits that the accumulated time series data exhibit exponential variation characteristics, a property that can be mathematically substantiated [21]. Leveraging this theory, a yield data-driven model can be formulated to capture the underlying trends and dynamics of the data. The model development typically involves the following steps:

$$\left[Q^{(2)}(t)\right]' = aQ^{(2)}(t) + BU^{(2)}(t) \tag{4}$$

where $\left[Q^{(2)}(t)\right]'$ represents the derivative with respect to time t; $U^{(2)}(t) = \begin{pmatrix} U_1^{(2)}(t) \\ U_2^{(2)}(t) \\ \cdots \\ U_3^{(2)}(t) \end{pmatrix}$; a and B are the non-identified parameters, and B is shown as follows:

$$B = (B_1, B_2, B_3, \ldots, B_7)$$

The above output data-driven Equation (4) is discretized by first-order approximation over time as follows:

$$Q_{k+1}^{(2)} = aQ^{(2)}(t) + BU^{(2)}(t) \tag{5}$$

Then, this Equation can be converted to the following equation:

$$Q_{k+1}^{(2)} = aQ_k^{(2)} + B_1U_{1k+1}^{(2)} + B_2U_{2k+1}^{(2)} + \ldots + B_7U_{7k+1}^{(2)} \tag{6}$$

The derived parameters a and B, along with the flowchart detailing the least squares fitting process, are depicted in Figure 2. The dimensionless data, following the second accumulation, were fed into Equation (6) for comprehensive multiple linear regression analysis. Employing the least squares method, we calculated the sum of squared residuals between the actual observed values and those predicted by the model. This approach was instrumental in determining the optimal parameter values, thereby enhancing the model's predictive accuracy and reliability.

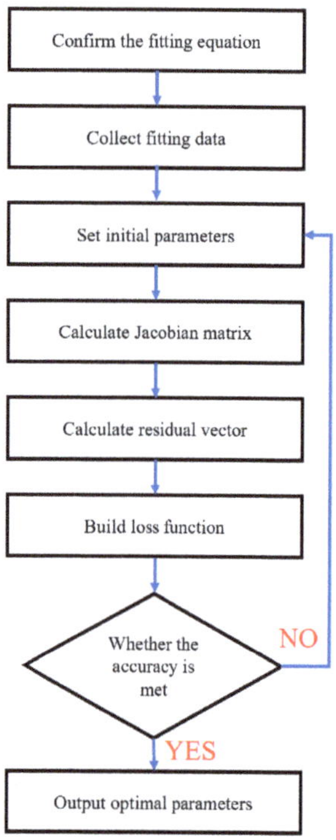

Figure 2. Flow chart of the least squares fitting.

2.4. Dimension Recovery

The predicted gas production is derived after two subtractions, utilizing the fitted equation. Specifically, the model is applied with the output data from year $k+1$ and the main control factors for year $k+1$. Thereafter, the dimensionless production data for year $k+1$, after two accumulations, are computed. Ultimately, the predicted gas production is ascertained through a series of two subtractions and dimension recovery processes, thereby yielding a precise forecast.

The first subtraction can be calculated as follows:

$$Q_{k+1}^{(1)} = Q_{k+1}^{(2)} - Q_k^{(2)} \tag{7}$$

where $Q_{k+1}^{(1)}$ is the production dimensionless data in $k+1$ after the first accumulation; $Q_{k+1}^{(2)}$ is the production dimensionless data in k after the second accumulation; and $Q_k^{(2)}$ is the production dimensionless data in k after the second accumulation.

The second subtraction can be calculated as follows:

$$Q_{k+1}^{(0)} = Q_{k+1}^{(1)} - Q_k^{(1)} \tag{8}$$

where $Q_{k+1}^{(0)}$ is the production dimensionless data in $k+1$; $Q_k^{(1)}$ is the production dimensionless data in k after the first accumulation.

The predicted gas production can be obtained as follows:

$$Q_{k+1} = Q_{k+1}^{(0)} \sum_1^n Q_n \tag{9}$$

where Q_{k+1} is the production data in $k+1$; $\sum_1^n Q_n$ is the sum of the gas production data.

3. Field Example

3.1. Data Preprocessing

The lithology of the targeted volcanic gas reservoir predominantly comprises acid tuff breccias, rhyolite, and andesite. The internal interlayering within the gas-bearing strata of the main rock mass is underdeveloped, with an interlayer density of merely 0.03 m/m, and the individual gas layers exhibit substantial thickness. A tuffaceous breccia septum, approximately 13 m thick, is present in the upper section of the gas layer, while a more substantial septum, about 170 m thick, is observed in the lower part. The fractures within the lower septum are relatively well developed, constituting 50.5% of the septum's total thickness. The porosity within the reservoir fluctuates between 7.1% and 22.2%, with an average value of 14.4%. The permeability varies widely, from 0.005 mD to 836.000 mD, averaging at 0.844 mD, indicating a heterogeneous subsurface environment.

The data indices for the targeted volcanic rock well area were categorized into two distinct groups: the prediction index, which represents gas production, and the main control factor index, encompassing the number of wells in operation, production time, effective thickness of the reservoir, controlled reserves, reserve abundance, formation pressure, and energy storage factor. Subsequently, all index data were meticulously organized in annual chronological order, as presented in Table 1.

Table 1. Annual natural gas production and main control factor index table.

Time /Year	$Q(t)$ /$\times 10^8$ m^3	$U_1(t)$ /Well	$U_2(t)$ /$\times 10^3$ h	$U_3(t)$ /m	$U_4(t)$ /$\times 10^8$ m^3	$U_5(t)$ /$\times 10^8$ m^3/km^2	$U_6(t)$ /MPa	$U_7(t)$ /-
2008	0.05	3	0.89	91.30	18.4	15.51	46.61	9.43
2009	1.15	8	26.65	94.64	22.6	15.41	45.05	9.46
2010	1.94	11	50.49	86.99	25.2	15.25	43.49	8.81
2011	1.96	13	57.51	80.92	32.3	15.09	41.92	8.31
2012	1.77	15	68.45	81.56	36.0	14.94	40.36	7.83
2013	2.27	17	85.78	86.77	43.3	14.75	38.80	9.01
2014	2.35	15	86.99	91.13	69.3	14.56	37.30	9.83
2015	2.71	17	96.56	89.47	77.1	14.33	35.90	9.61
2016	2.5	20	98.74	89.93	82.9	14.12	34.50	9.95
2017	2.66	21	107.23	86.47	90.1	13.90	33.10	9.38
2018	2.39	21	97.42	87.00	95.7	13.70	31.80	9.50

In accordance with Equation (1), the original data presented in Table 1 were subjected to dimensionless processing, and the outcomes are detailed in Table 2. The distribution of these dimensionless data points is graphically represented in Figure 3. The maximum values of the dimensionless data range from 0.089 to 0.144, indicating a variation of 38.19%. Conversely, the minimum values span from 0.001 to 0.077, reflecting a substantial change of 98.70%. It is evident that there are considerable disparities among the various data indices, highlighting the importance of normalization in comparative analyses.

Table 2. Data after dimensionless processing.

Time/Year	$Q^0(t)$	$u_1^0(t)$	$u_2^0(t)$	$u_3^0(t)$	$u_4^0(t)$	$u_5^0(t)$	$u_6^0(t)$	$u_7^0(t)$
2008	0.002	0.017	0.001	0.087	0.027	0.089	0.101	0.085
2009	0.047	0.044	0.030	0.090	0.033	0.088	0.098	0.086
2010	0.079	0.061	0.057	0.083	0.036	0.087	0.095	0.080
2011	0.080	0.072	0.064	0.077	0.047	0.086	0.091	0.075
2012	0.072	0.083	0.077	0.077	0.052	0.085	0.088	0.071
2013	0.093	0.094	0.096	0.082	0.062	0.084	0.084	0.081
2014	0.096	0.083	0.097	0.087	0.100	0.083	0.081	0.089
2015	0.111	0.094	0.108	0.085	0.111	0.082	0.078	0.087
2016	0.102	0.110	0.110	0.085	0.120	0.081	0.075	0.090
2017	0.109	0.116	0.120	0.082	0.130	0.079	0.072	0.085
2018	0.098	0.116	0.109	0.083	0.138	0.078	0.069	0.086
2019	0.110	0.110	0.131	0.083	0.144	0.077	0.066	0.086

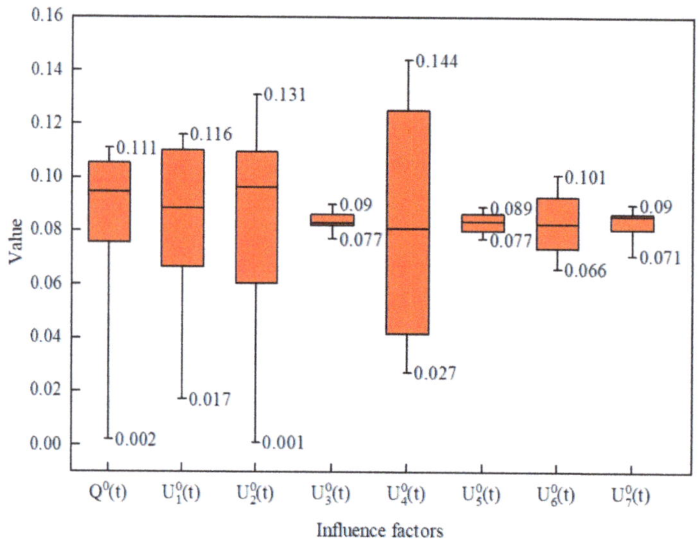

Figure 3. The distribution of the dimensionless data.

Utilizing Equation (2), the original data were subjected to the first accumulation, with the outcomes documented in Table 3. The distribution of the dimensionless data after the first accumulation is depicted in Figure 4. Notably, the maximum values of the dimensionless data after this initial accumulation uniformly reach 1, signifying no variation. The minimum values, on the other hand, range from 0.001 to 0.101, representing a reduction of 99.01%. This observation underscores that the variability in the dimensionless data following the first accumulation is markedly diminished compared to the initial dimensionless data set, thereby illustrating the effectiveness of the accumulation process in stabilizing data fluctuations.

In advance of the modeling phase, the dimensionless data following the first accumulation were further accumulated using Equation (3), with the results presented in Table 4. The distribution of the dimensionless data subsequent to the second accumulation is illustrated in Figure 5. The maximum values of the dimensionless data after this second accumulation range from 0.089 to 0.144, demonstrating a variation of 38.19%. The minimum values span from 0.001 to 0.077, indicating a substantial change of 98.70%. These refined data values, exhibiting reduced variability, are now suitable for inclusion in the modeling process.

Table 3. Data after the first cumulative processing.

Time/Year	$Q^1(t)$	$u_1^1(t)$	$u_2^1(t)$	$u_3^1(t)$	$u_4^1(t)$	$u_5^1(t)$	$u_6^1(t)$	$u_7^1(t)$
2008	0.002	0.017	0.001	0.087	0.027	0.089	0.101	0.085
2009	0.049	0.061	0.031	0.177	0.059	0.177	0.200	0.171
2010	0.128	0.122	0.087	0.259	0.096	0.264	0.294	0.250
2011	0.209	0.193	0.152	0.336	0.142	0.350	0.385	0.326
2012	0.281	0.276	0.228	0.413	0.194	0.435	0.473	0.396
2013	0.374	0.370	0.324	0.496	0.257	0.519	0.558	0.478
2014	0.470	0.453	0.422	0.582	0.357	0.603	0.639	0.567
2015	0.581	0.547	0.530	0.667	0.468	0.684	0.717	0.654
2016	0.683	0.657	0.640	0.753	0.588	0.765	0.792	0.744
2017	0.792	0.773	0.760	0.835	0.718	0.844	0.864	0.829
2018	0.890	0.890	0.869	0.917	0.856	0.923	0.934	0.914
2019	1.000	1.000	1.000	1.000	1.000	1.000	1.000	1.000

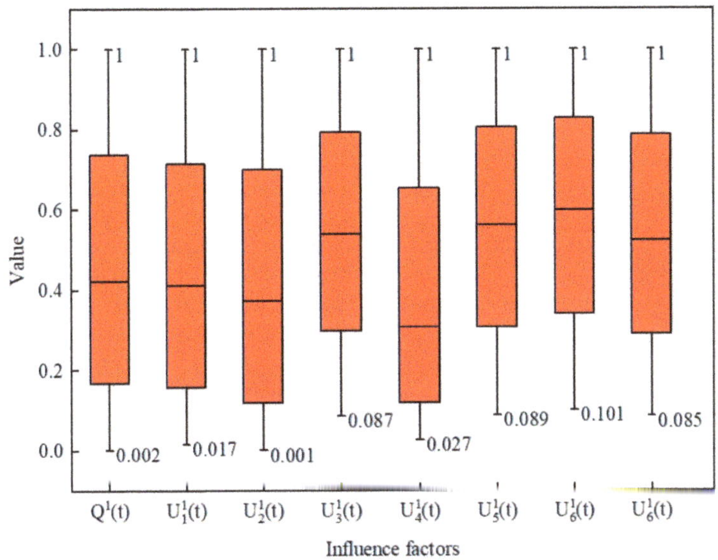

Figure 4. The distribution of the dimensionless data after the first accumulation.

Table 4. Data after the second cumulative processing.

Time/Year	$Q^2(t)$	$u_1^2(t)$	$u_2^2(t)$	$u_3^2(t)$	$u_4^2(t)$	$u_5^2(t)$	$u_6^2(t)$	$u_7^2(t)$
2008	0.002	0.017	0.001	0.087	0.027	0.089	0.101	0.085
2009	0.051	0.077	0.032	0.263	0.086	0.265	0.301	0.256
2010	0.180	0.199	0.119	0.522	0.181	0.529	0.595	0.507
2011	0.388	0.392	0.271	0.858	0.323	0.879	0.981	0.832
2012	0.669	0.669	0.499	1.272	0.518	1.314	1.454	1.229
2013	1.043	1.039	0.823	1.768	0.774	1.833	2.012	1.707
2014	1.514	1.492	1.245	2.350	1.131	2.436	2.651	2.273
2015	2.095	2.039	1.775	3.017	1.599	3.120	3.368	2.927
2016	2.778	2.696	2.415	3.770	2.186	3.886	4.160	3.671
2017	3.570	3.470	3.175	4.605	2.904	4.730	5.025	4.499
2018	4.460	4.359	4.044	5.522	3.759	5.653	5.958	5.414

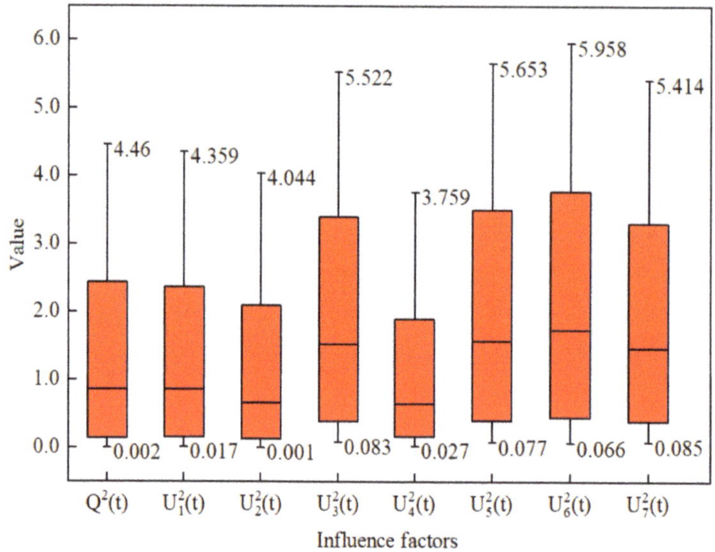

Figure 5. The distribution of the dimensionless data after the second accumulation.

3.2. Modeling

The dimensionless data after the second accumulation are input into Equation (6), then the a and B parameters are obtained after multiple linear regression simulation training by using the least squares method, and the results are shown below:

$$\begin{aligned} Q_{k+1}^{(2)} = -0.514Q_k^{(2)} + 0.147{U_1^{(2)}}_{k+1} + 1.631{U_2^{(2)}}_{k+1} - 14.301{U_3^{(2)}}_{k+1} \\ +0.266{U_4^{(2)}}_{k+1} + 0.441{U_5^{(2)}}_{k+1} + 4.995{U_6^{(2)}}_{k+1} \\ +8.314{U_7^{(2)}}_{k+1} \end{aligned} \tag{10}$$

The gas production data and predicted values for the years 2008 to 2019, along with the projected production for 2020 to 2024, are graphically represented in Figure 6. A clear observation is that the predicted production aligns closely with the actual production across different years, indicating that the gas field reached a stable production phase by 2019. Additionally, Figure 7 illustrates the discrepancies between the actual and predicted gas production. The prediction error ranges from 0.46% to 30.47%, with an average error of 12.49%. It is evident that the error tends to diminish progressively with the extension of the mining period and the accumulation of production data, suggesting an improvement in the predictive accuracy over time.

3.3. Modeling Evaluation

3.3.1. Coefficient of Determination

The coefficient of determination (R^2) is a commonly used statistical indicator to evaluate the fit between the regression model and observed data [14–16]. It obviously indicates that the fitting model can explain the proportion of observed data variance, with a range of values between 0 and 1. The higher the R^2, the better the fit, and it can be calculated as follows:

$$\tilde{y} = \frac{1}{n}\sum_{i=1}^{n} y_i \tag{11}$$

$$SST = \sum_i \left(y_i - \tilde{y}\right)^2 \tag{12}$$

$$SSR = \sum_i (y_i - y_i')^2 \tag{13}$$

$$R^2 = 1 - \frac{SSR}{SST} \tag{14}$$

where $\widetilde{y}$ is the average of observed value; y_i' is the predicted value; y_i is the true value; SST is the sum of squares in the real data; and SSR is the sum of residual squares.

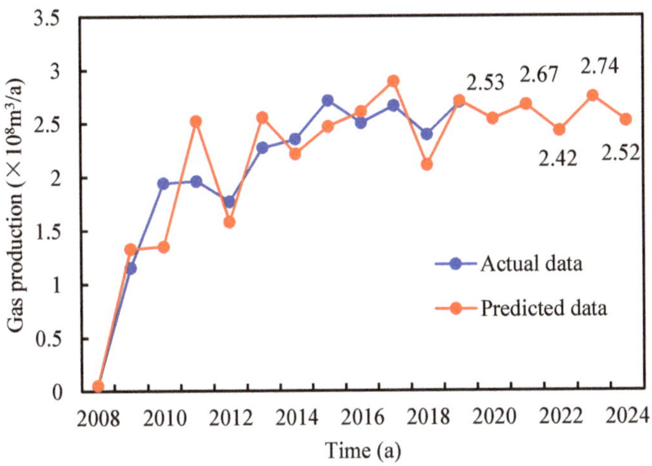

Figure 6. The gas production and predicted data.

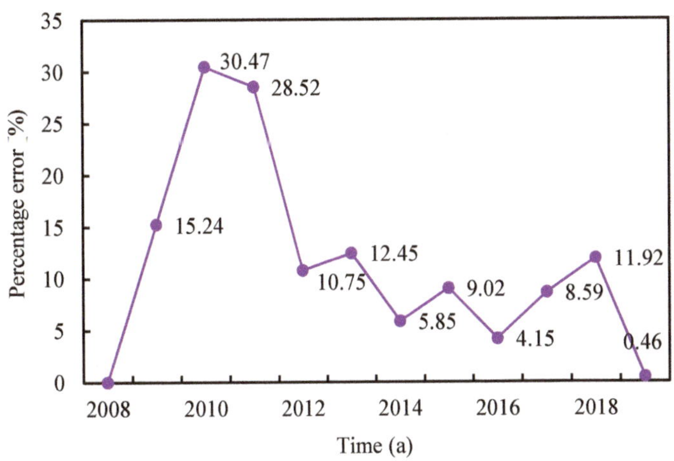

Figure 7. The error between gas production and predicted data.

Utilizing the gas production and predicted data from 2009 to 2019, the coefficient of determination was calculated to be 0.99, as per Equations (11)–(14). This high value for the coefficient of determination signifies that the gas production data are closely aligned with the predictions of the fitting model, thereby confirming the model's efficacy in capturing the production trends over the specified period.

3.3.2. Significance Testing

According to the relevant data of the established multiple linear regression fitting model, the significance of the regression equation is tested [22–25].

Hypothesis: $H_{0i}: \beta_i = 0, i\epsilon\{1, 2, \cdots, 12\}$, the sum of squared deviations are calculated as follows:

$$S_T^2 = \sum_i^n (y_i - \bar{y}) \quad (15)$$

$$S_R^2 = \sum_i^n (\hat{y} - \bar{y}) \quad (16)$$

$$S_E^2 = S_T^2 - S_R^2 \quad (17)$$

$$F = \frac{S_R^2/12}{S_R^2/12} \quad (18)$$

F-value is a statistical measure of analysis of variance used to test whether the regression equation is significant [26–30]. Taking a significance level of $F = 0.05$, the distribution table shows $F_{1-\alpha}(k, n-k-1) = F_{0.95}(729, 1974) = 1 < 19.249$; thus, the hypothesis should be rejected.

The F-values for $Q(t)$ (gas production), $U_1(t)$ (number of wells in operation), $U_2(t)$ (production time), $U_3(t)$ (effective thickness of reservoir), $U_4(t)$ (controlled reserves), $U_5(t)$ (reserve abundance), $U_6(t)$ (formation pressure), and $U_7(t)$ (energy storage factor) are graphically represented in Figure 8. A clear trend is evident, with the F-values arranged in descending order as follows: $U_2(t)$, $U_3(t)$, $U_7(t)$, $U_6(t)$, $Q(t)$, $U_4(t)$, $U_1(t)$, and $U_5(t)$. This ranking underscores the increasing influence of the annual production time, effective thickness of the reservoir, energy storage factor, formation pressure, gas production, controlled reserves, number of wells in operation, and reserve abundance on the annual gas production. The visualization of these F-values provides a quantitative assessment of the relative significance of each factor in the predictive model [31].

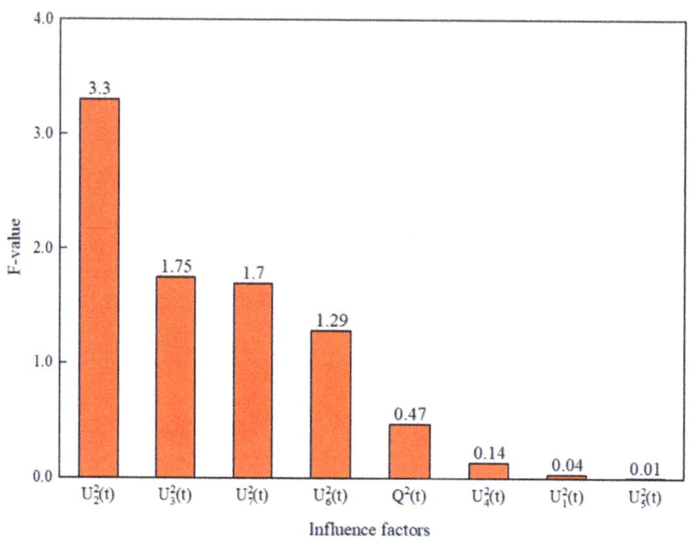

Figure 8. F-values of various influencing factors.

4. Conclusions

In this study, standardization ensures data consistency and comparability, enhancing data quality for analysis. Cumulative operations reduce random fluctuations, highlighting long-term trends for better model predictions. The application of two accumulations of dimensionless gas production data effectively mitigates the inherent randomness of historical data, thereby establishing a robust fitting sample library.

Based on on-site construction experience, considering the time-varying characteristics of the gas well quantity, production time, effective reservoir thickness, controlled reserves, reserve abundance, formation pressure, and energy storage coefficient, a data-driven method was used to establish a natural gas production prediction model based on differential simulation theory. The calculation results showed that the average error between the actual production and predicted production was 12.49%, and the model determination coefficient was 0.99, indicating that the model can effectively predict natural gas production.

The *F*-values in descending order are $U_2(t)$, $U_3(t)$, $U_7(t)$, $U_6(t)$, $Q(t)$, $U_4(t)$, $U_1(t)$, and $U_5(t)$, indicating that the influence of the reserve abundance, number of wells in operation, controlled reserves, previous year's gas production, formation pressure, energy storage coefficient, effective thickness of matrix, and annual production time on the annual gas production gradually increases.

Author Contributions: Conceptualization, H.Z.; methodology, J.P.; software, L.Z.; validation, H.D.; formal analysis, H.D.; investigation, X.T.; resources, X.M.; data curation, Y.X.; writing—original draft preparation, J.Y.; writing—review and editing, Z.L.; visualization, Y.X.; supervision, J.Y.; project administration, Z.L. All authors have read and agreed to the published version of the manuscript.

Funding: This research was supported by the Science and Technology Research Program of Chongqing Municipal Education Commission (Grant No. KJZD-M202201502).

Data Availability Statement: The original contributions presented in the study are included in the article, further inquiries can be directed to the corresponding author.

Conflicts of Interest: Authors Haijie Zhang, Junwei Pu, Li Zhang and Hengjian Deng were employed by the company Chong Qing Shale Gas Exploration and Development Company Limited; Author Yingming Xie was employed by the company CNOOC EnerTech-Drilling & Production Co.; Authors Xiaochang Tong, Xiangjie Man and Zhonghua Liu were employed by the company Pancasia Holding Company of Limited Liability. The authors declare that the research was conducted in the absence of any commercial or financial relationships that could be construed as a potential conflict of interest.

References

1. Rahmanifard, H.; Gates, I. A Comprehensive review of data-driven approaches for forecasting production from unconventional reservoirs: Best practices and future directions. *Artif. Intell. Rev.* **2024**, *57*, 213. [CrossRef] [PubMed]
2. Liu, Z.; Bai, B.; Ding, Z.; Qu, H.; Zeng, S.; Da, X. Impact of cleanup additive on methane desorption on Longmaxi shale. *Fuel* **2021**, *300*, 121003. [CrossRef]
3. Syed, F.I.; Alnaqbi, S.; Muther, T.; Dahaghi, A.K.; Negahban, S. Smart shale gas production performance analysis using machine learning applications. *Pet. Res.* **2022**, *7*, 21–31. Available online: https://doaj.org/article/cb883d66b51743e89790c0462cd50902 (accessed on 22 August 2024). [CrossRef]
4. Mishra, S.; Lin, L. Application of data analytics for production optimization in unconventional reservoirs: A critical review. In Proceedings of the SPE/AAPG/SEG Unconventional Resources Technology Conference, Austin, TX, USA, 24–26 July 2017. [CrossRef]
5. Mohaghegh, S.D. Reservoir modeling of shale formations. *J. Nat. Gas Sci. Eng.* **2013**, *12*, 22–33. [CrossRef]
6. Li, D.; You, S.; Liao, Q.; Sheng, M.; Tian, S. Prediction of Shale Gas Production by Hydraulic Fracturing in Changning Area Using Machine Learning Algorithms. *Transp. Porous Med.* **2023**, *149*, 373–388. [CrossRef]
7. Mohammadmoradi, P.; Moradi, H.M.; Kantzas, A. Data-driven production forecasting of unconventional wells with apache spark. In Proceedings of the SPE Western Regional Meeting, Garden Grove, CA, USA, 22–26 April 2018. [CrossRef]
8. Enyioha, C.; Ertekin, T. Performance prediction for advanced well structures in unconventional oil and gas reservoirs using artificial intelligent expert systems. In Proceedings of the SPE Annual Technical Conference and Exhibition, San Antonio, TX, USA, 9–11 October 2017. [CrossRef]

9. Wang, S.; Chen, S. Insights to fracture stimulation design in unconventional reservoirs based on machine learning modeling. *J. Pet. Sci. Eng.* **2019**, *174*, 682–695. [CrossRef]
10. Han, D.; Jung, J.; Kwon, S. Comparative study on supervised learning models for productivity forecasting of shale reservoirs based on a data-driven approach. *Appl. Sci.* **2020**, *10*, 1267. [CrossRef]
11. Han, D.; Kwon, S.; Son, H.; Lee, J. Production forecasting for shale gas well in transient flow using machine learning and decline curve analysis. In Proceedings of the Asia Pacific Unconventional Resources Technology Conference, Brisbane, Australia, 18–19 November 2019. [CrossRef]
12. Liao, L.; Zeng, Y.; Liang, Y.; Zhang, H. Data mining: A novel strategy for production forecast in tight hydrocarbon resource in Canada by random forest analysis. In Proceedings of the International Petroleum Technology Conference, Dhahran, Kingdom of Saudi Arabia, 13–15 January 2020. [CrossRef]
13. Bhattacharyya, S.; Vyas, A. Application of machine learning in predicting oil rate decline for Bakken shale oil wells. *Sci. Rep.* **2022**, *12*, 16154. [CrossRef]
14. Zhou, Q.; Dilmore, R.; Kleit, A.; Wang, J.Y. Evaluating gas production performances in marcellus using data mining technologies. In Proceedings of the Unconventional Resources Technology Conference, Denver, CO, USA, 25–27 August 2014; pp. 20–36. [CrossRef]
15. Grujic, O.; Da Silva, C.; Caers, J. Functional approach to data mining, forecasting, and uncertainty quantification in unconventional reservoirs. In Proceedings of the SPE Annual Technical Conference and Exhibition, Houston, Texas, USA, 28–30 September 2015; pp. 1704–1720. [CrossRef]
16. Lolon, E.; Hamidieh, K.; Weijers, L.; Mayerhofer, M.; Melcher, H.; Oduba, O. Evaluating the relationship between well parameters and production using multivariate statistical models: A middle Bakken and three forks case history. In Proceedings of the SPE Hydraulic Fracturing Technology Conference, The Woodlands, TX, USA, 9–11 February 2016. [CrossRef]
17. Khanal, A.; Khoshghadam, M.; Lee, W.J.; Nikolaou, M. New forecasting method for liquid rich shale gas condensate reservoirs with data driven approach using principal component analysis. *J. Nat. Gas Sci. Eng.* **2017**, *38*, 621–637. [CrossRef]
18. Xue, L.; Liu, Y.; Xiong, Y.; Liu, Y.; Cui, X.; Lei, G. A data-driven shale gas production forecasting method based on the multi-objective random forest regression. *J. Pet. Sci. Eng.* **2021**, *196*, 107801. [CrossRef]
19. Johan, D.C.; Shukla, P.; Shrivastava, K.; Koley, M. Data-Driven Completion Optimization for Unconventional Assets. In Proceedings of the Unconventional Resources Technology Conference, Denver, CO, USA, 13–15 June 2023; p. 3504. [CrossRef]
20. Ghadami, A.; Epureanu, B.I. Data-driven prediction in dynamical systems: Recent developments. *Philos. Trans. R. Soc. A Math. Phys. Eng. Sci.* **2022**, *380*, 1429–1442. [CrossRef] [PubMed]
21. Mohaghegh, S.D. *Shale Analytics: Data-Driven Analytics in Unconventional Resources*, 1st ed.; Springer: Cham, Switerland, 2017.
22. Ma, H.; Zhao, W.; Zhao, Y.; He, Y. A data-driven oil production prediction method based on the gradient boosting decision tree regression. *Comput. Model. Eng. Sci.* **2023**, *134*, 1773–1790. [CrossRef]
23. De Oliveira Werneck, R.; Prates, R.; Moura, R.; Goncalves, M.M.; Castro, M.; Soriano-Vargas, A.; Junior, P.R.; Hossain, M.M.; Zampieri, M.F.; Ferreira, A.; et al. Data-driven deep-learning forecasting for oil production and pressure. *J. Pet. Sci. Eng.* **2022**, *210*, 109937. [CrossRef]
24. Gryzlov, A.; Safonov, S.; Arsalan, M. Intelligent Production Monitoring with Continuous Deep Learning Models. *SPE J.* **2022**, *27*, 1304–1320. [CrossRef]
25. Chi, L.; Su, H.; Zio, E.; Qadrdan, M.; Li, X.; Zhang, L.; Fan, L.; Zhou, J.; Yang, Z.; Zhang, J. Data-driven reliability assessment method of Integrated Energy Systems based on probabilistic deep learning and Gaussian mixture Model-Hidden Markov Model. *Renew. Energy* **2021**, *174*, 952–970. [CrossRef]
26. Chen, D.; Huang, C.; Wei, M. Shale Gas Production Prediction Based on PCA-PSO-LSTM Combination Model. *J. Circuits Syst. Comput.* **2024**, *33*, 2450176. [CrossRef]
27. He, Y.W.; He, Z.Y.; Tang, Y.; Xu, Y.J.; Long, J.C.; Sepehrnoori, K. Shale gas production evaluation framework based on data-driven models. *Pet. Sci.* **2023**, *20*, 1659–1675. [CrossRef]
28. Meng, J.; Zhou, Y.J.; Ye, T.R.; Xiao, Y.T.; Lu, Y.Q.; Zheng, A.W.; Liang, B. Hybrid data-driven framework for shale gas production performance analysis via game theory, machine learning, and optimization approaches. *Pet. Sci.* **2023**, *20*, 277–294. [CrossRef]
29. Wang, Y.; Wang, Z.; Qian, S.; Qiu, X.; Shen, W.; Zhang, X.; Lyu, B.; Cui, J. Data-driven framework for prediction and optimization of gas turbine blade film cooling. *Phys. Fluids* **2024**, *36*, 035160. [CrossRef]
30. Brilliant, L.S.; Dulkarnaev, M.R.; Danko, M.Y.; Lisheva, A.O.; Nabiev, D.K.; Khutornaya, A.I.; Malkov, I.N. Oil production management based on neural network optimization of well operation at the pilot project site of the Vatyeganskoe field (Territorial Production Enterprise Povkhneftegaz). *Georesursy* **2022**, *24*, 3–15. [CrossRef]
31. Bahaloo, S.; Mehrizadeh, M.; Najafi-Marghmaleki, A. Review of application of artificial intelligence techniques in petroleum operations. *Pet. Res.* **2023**, *8*, 167–182. [CrossRef]

Disclaimer/Publisher's Note: The statements, opinions and data contained in all publications are solely those of the individual author(s) and contributor(s) and not of MDPI and/or the editor(s). MDPI and/or the editor(s) disclaim responsibility for any injury to people or property resulting from any ideas, methods, instructions or products referred to in the content.

energies

MDPI

Article

Development of a Hybrid AI Model for Fault Prediction in Rod Pumping System for Petroleum Well Production

Aoxue Zhang [1], Yanlong Zhao [1,*], Xuanxuan Li [1], Xu Fan [1], Xiaoqing Ren [1], Qingxia Li [2] and Leishu Yue [1]

[1] School of Petroleum, China University of Petroleum-Beijing at Karamay, Karamay 834000, China; 2022216760@st.cupk.edu.cn (A.Z.); 2022216758@st.cupk.edu.cn (X.L.); 2022216757@st.cupk.edu.cn (X.F.); 2022216759@st.cupk.edu.cn (X.R.); 2023216807@st.cupk.edu.cn (L.Y.)

[2] Region of Luliang Oilfield, PetroChina Xinjiang Oilfield Company, Karamay 834000, China; lqxll2014@petrochina.com.cn

* Correspondence: zhaoyanlong@cupk.edu.cn; Tel.: +86-18010092639

Abstract: Rod pumping systems are widely used in oil wells. Accurate fault prediction could reduce equipment fault rate and has practical significance in improving oilfield production efficiency. This paper analyzed the production journal of rod pumping wells in block X of Xinjiang Oilfield. According to the production journal, oil well maintenance operations are primarily caused by five types of faults: scale, wax, corrosion, fatigue, and wear. These faults make up approximately 90% of all faults. 1354 oil wells in the block that experienced workover operations as a result of the aforementioned factors were chosen as the research objects for this paper. To lower the percentage of data noise, wavelet threshold denoising and variational mode decomposition were used. Based on the bidirectional long short-term memory network, an intelligent model for fault prediction was built. It was trained and verified with the help of the sparrow search algorithm. Its efficacy was demonstrated by testing various deep learning models in the same setting and with identical parameters. The results show that the prediction accuracy of the model is the highest compared with other 11 models, reaching 98.61%. It is suggested that the model using artificial intelligence can provide an accurate fault warning for the oilfield and offer guidance for the maintenance of the rod pumping system, which is meant to reduce the occurrence of production stagnation and resource waste.

Keywords: rod pumping system; oilfield; fault prediction; artificial intelligence; fault causes

1. Introduction

The rod pumping system has historically been dominant in oilfield production [1]. The system moves up and down through the rod connected by the motor to transfer the surface energy to the plunger pump and the downhole fluid so as to produce oil. However, due to the complex environmental conditions, coupled with the influence of sand, water, gas, and other external factors [2,3], all the rod pumping wells are subject to downtime and faults, including scale, wax, corrosion and partial wear of sucker rods and tubing, fatigue, etc. Long-term maintenance is deemed essential. Nearly USD 12,000 will be spent for a workover operation. It costs oilfields a great amount of labor and money to overhaul rod pumping wells every year and this has a negative effect on production efficiency and economic benefits [4]. If oilfields were able to predict the working condition of rod pumping system, chemical inhibitors and parameters in the conditions of the production would be adjusted as precautionary measures. Therefore, fault prediction of rod pumping system in a timely and accurate manner means a lot to oil production [5,6].

So far, there are two main types of methods for fault prediction in the rod pumping system: one is the physics-based model, and the other is the data-driven model, which regards big data as the main drive [7]. Table 1 shows the evolution of fault diagnosis technology in the rod pumping system, guiding the fault prediction technique. In the early stages, the traditional method of diagnosing faults in the rod pumping system was to analyze and judge the downhole conditions by experts. It means that the accuracy rate depends on the experience of experts. Moreover, it could not grasp the downhole conditions in real time. In the late 1920s, the surface dynamometer card based on the movement of polished rod was presented, which allowed the operator to assess the working condition of the wells. In the 1930s, Gilbert [8] invented the downhole dynamometer, which was sent into the downhole alongside pumps. It was directly used to measure the pump indicator cards to obtain the real production data of the pump. In the 1960s, Gibbs and Neely [9] put forward a computer diagnostic technique for the rod pumping system. According to the propagation process of force, the wave equation was established and solved to obtain the indicator cards. It could solve the equation of the indicator diagram but still rely on manual analysis. In the late 1980s, a rule-based expert system for rod pump unit diagnostics was developed by Derek et al. [10]. And then Foley and Svinos [11], and Schirmer et al. [12] upgraded the expert system for rod-pumping diagnosis. In the 1990s, Rogers et al. [13] introduced artificial neural networks (ANNs) into the field of condition monitoring in oil wells for the first time. Ashenayi et al. [14] trained the multilayer neural network model to identify significant features of the surface dynamometer cards containing 11 underground working conditions. After entering the 21st century, with the development of artificial intelligence technology, various fault diagnosis methods emerge endlessly and gradually move toward automation and intelligence.

Table 1. Development of Fault Diagnosis in Rod Pumping System.

Age	Development
1920s	The surface dynamometer card based on the movement of polished rod was presented, allowing the operator to assess the working condition of the wells.
1930s–1950s	The downhole dynamometer was invented and was directly used to measure the pump indicator cards to obtain the real production data of the pump.
1960s–1970s	The computer diagnostic technique for the rod pumping system was put forward. It obtained the indicator cards by solving the wave equation.
1980s	A rule-based expert system for rod pump unit diagnostics was developed.
1990s	ANN was introduced into the field of condition monitoring in oil wells for the first time.
2000s	More artificial neural networks were used for the recognition of dynamometer cards. Various software with correlation to condition monitoring were developed.
2010s	Combined with artificial intelligence technology, the fault diagnosis methods for rod pumping were moving towards intelligence.
2020s	With the maturity of industrial Internet technology, the accuracy of fault diagnosis is further improved.

Generally, such physics-based models are exploring the physics behind the technical topic in good means, however, they are suffering from limitations in terms of model applications, limited database, low dimensionality, and low accuracy [15]. In recent years, the emergence of new technologies, such as big data and deep learning, has promoted the establishment of Smart Oilfield and provided a new approach to the problem of intelligent fault prediction in rod pumping systems.

According to the research status of domestic and foreign scholars on the fault prediction of rod pumping systems, the fault prediction based on dynamometer cards is the most popular. Gao et al. [16] (2015) constructed an extreme learning machine (ELM) associated with the characteristics of the dynamometer card for diagnosing faults in rod pump units. Lv et al. [17] (2021) proposed an evolutionary support vector machine (SVM) method based on incremental algorithm and simulated indicator diagrams for working condition detection in sucker rod pumping systems. Yin et al. [18] (2023) used a mini-batch method to improve conditional generative adversarial networks (CGANs) to solve the problem of imbalanced working states recognition. He et al. [19] (2024) recognized the working condition using 4-segment time-frequency signature matrix (4S-TFSM) and deep learning.

Many experts and scholars have conducted extensive research on fault prediction methods based on electrical parameters. Zheng et al. [20] (2019) used electric power curves to extract statistical features and used hidden Markov models to detect faults for sucker rod pumps. Wei and Gao [21] (2020) proposed a new method called deep and breadth learning system (DBLS), which is a new method for rod-pumping fault detection based on motor power. Bai et al. [22] (2022) put forward a method for generating electrical parameter samples based on time series transform inversion power generation reverse network (TSC-DCGAN) to address the issue of insufficient data samples for electrical parameters in rod pumping. Liu et al. [23] (2024) proposed a method based on northern goshawk algorithm optimized variational mode decomposition (NGO-VMD) and least squares vector machine (LSSVM).

Furthermore, the production journal is also one of the most important parts of big data in the oilfield besides dynamometer cards and electrical parameters, yet it is not commonly used in fault prediction.

This paper analyzes the production journal of various sucker rod pumping wells in block X of Xinjiang Oilfield. Variational mode decomposition (VMD) and wavelet threshold denoising were used to reduce the proportion of data noise. Sparrow search algorithm (SSA) was used to optimize the bidirectional long short-term memory (BiLSTM) network model. A neural network model that can predict the type of faults to happen was trained, hoping to provide fault warning accurately and guide the maintenance.

2. Oilfield Overview and Data Collection

The data in this paper are from block X of Xinjiang Oilfield. The beam pumping unit is the main type of pumping units used in block X, accounting for more than 90%. The reservoir in the block can be described as low oil-saturation. The block has 1484 wells with an average pump setting depth of 1430 m. The present daily production is 18,600 bbl of oil. Its average pump detection period is approximately 330 days.

On the one hand, the average well depth of the oilfield is relatively deep. Due to the irregular wellbores formed in the drilling process, the rods were easily destabilized and bent during the reciprocating movement, resulting in the rod and tube partial wear. On the other hand, at present, block X has entered the water-cut rising period. With the extension of the production period of oil, the corrosive output liquid contacted the pipe body and led to the corrosion of the pipe. Fatigue crack and corrosion promoted each other, resulting in rod fatigue, tubing leakage, and other faults frequently.

The production journal indicates that five kinds of faults including scale, wax, corrosion, fatigue, and wear account for about 90% of all faults, which are the main reasons for oil well maintenance operations. The classification and statistics of five fault causes are shown in Figure 1 as follows: corrosion (26.2%), fatigue (7.2%), scale (9.5%), wax (23.7%), and wear (33.3%).

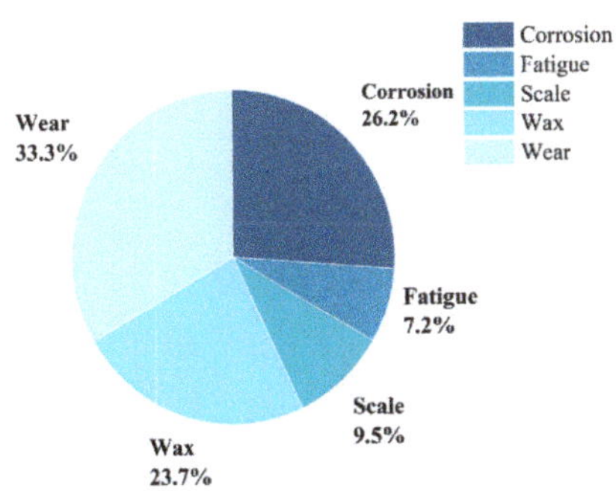

Figure 1. Proportion of five fault causes.

This paper selected 1354 oil wells in the block that caused workover operations due to the above five fault causes as the research objects. There are a total of 837,625 pieces of data, including chloride ion concentration, bicarbonate concentration, sulfate concentration, total salinity, formation crude oil density, formation crude oil viscosity, daily liquid production, daily oil production, water content, pump hanging depth, tubing pressure, casing pressure, stroke, and stroke times, a total of 14 kinds of characteristic information in all. The data were arranged in chronological order of production. These data record the changes of the pumping wells within a limited period of time, which can reflect the working state of rod pumping.

3. Methodologies

Variational mode decomposition (VMD) and wavelet threshold denoising were employed to reduce the proportion of data noise. Variational mode decomposition can decompose time series data into a series of intrinsic mode functions (IMF). Sparrow search algorithm (SSA) was used to optimize the bidirectional long short-term memory (BiLSTM) networks model. As a result, the intelligent model of fault prediction in the rod pumping system was constructed. The process of fault prediction based on VMD-SSA-BiLSTM is shown in Figure 2.

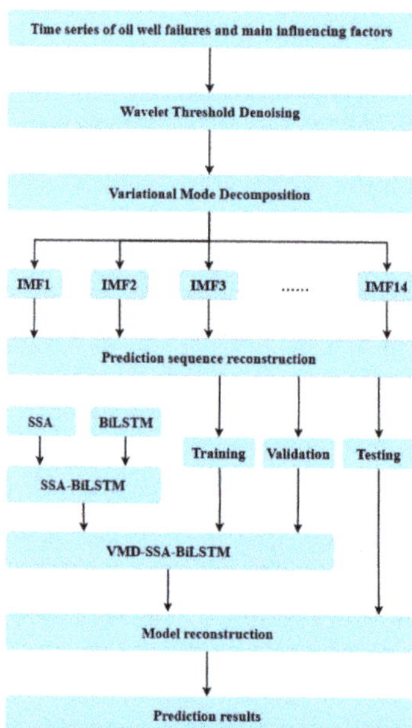

Figure 2. The structure diagram of the model.

3.1. Denoising Based on Variational Mode Decomposition

Variational mode decomposition (VMD) is an automatic adaptive, non-recursive information processing method proposed by Dragomiretskiy et al. [24]. VMD serves as a pivotal signal-processing method extensively applied in practical projects such as signal processing, data mining, and image processing. It can solve the problem of mode aliasing that occurs during the decomposition of raw signals. The main idea of this method is to generate and solve variational problems and achieve adaptive signal decomposition by determining the optimal solution of the variational model [25]. The VMD algorithm decomposes the data into K intrinsic mode function (IMF) components and corresponding K center frequencies. Its constraint model is as follows:

$$\begin{cases} \min\limits_{\{u_k,w_k\}} \{ \sum\limits_{k=1}^{K} \left\| \partial_t [(\delta(t) + \frac{j}{\pi t}) * u_k(t)] e^{-jw_k t} \right\|_2^2 \} \\ s.t. \sum\limits_{k}^{K} u_k(t) = f(t) \end{cases} \tag{1}$$

where ∂_t represents taking the partial derivative of t; $\delta(t)$ is the Dirac function; $*$ is the convolution operator; $u_k(t)$ is the kth modal component obtained through decomposition; $w_k(t)$ is the centrality corresponding to the kth modal component; and $f(t)$ is the original signal to be decomposed.

To transform the equation into an unconstrained optimization problem, an augmented Lagrangian function is introduced:

$$\begin{aligned} L(u_k, w_k, \lambda) = \alpha \sum_{k=1}^{K} \left\| \partial_t [(\delta(t) + \frac{j}{\pi t}) * u_k(t)] e^{-jw_k t} \right\|_2^2 + \\ \left\| f(t) - \sum_{k=1}^{K} u_k(t) \right\|_2^2 + \left\langle \lambda(t), f(t) - \sum_{k} u_k(t) \right\rangle \end{aligned} \tag{2}$$

where α is the quadratic penalty factor; λ is the Lagrange constant; and $\langle \cdot \rangle$ is the inner product operator.

As can be seen from the above, the VMD results are related to the number of modal components K and the penalty factor.

The advantage of VMD lies in its ability to specify the number of modes, avoiding mode aliasing and endpoint effects. It also ensures the sparsity of the decomposition results in the frequency domain, which can solve the problem of mode aliasing that occurs during the decomposition of the original signal. It performs excellently in decomposition accuracy, convergence speed, and anti-interference capabilities.

3.2. Wavelet Threshold Denoising

Wavelet threshold denoising requires selecting appropriate wavelet basis function and decomposition level based on the characteristics of the dataset, setting thresholds and threshold functions, processing wavelet coefficients, reconstructing signals, and reducing noise ratios to achieve the goal of highlighting effective information in the data [26].

The common threshold selection principles include rigrsure, minimaxi, sqtwolog, and heursure. This paper chose heursure. As a compromise between the rigrsure and sqtwolog, heuristic will determine whether to use rigrsure or sqtwolog based on the signal-to-noise ratio of the signal. It decides in the following way:

$$eta = \frac{\|x\|^2 - n}{n} \tag{3}$$

$$crit = \frac{\left[\frac{\log(n)}{\log 2}\right]^{1.5}}{\sqrt{n}} \tag{4}$$

$$\lambda = \begin{cases} \lambda_1, & eta < crit \\ \min(\lambda_1, \lambda_2), & eta \geq crit \end{cases} \tag{5}$$

where n is the length of the wavelet coefficient vector to be estimated; λ_1 is the threshold calculated by sqtwolog; and λ_2 is the threshold calculated by rigrsure.

According to relevant research and analysis [27–29], it was found that the denoising performance of soft thresholding is better than that of hard thresholding. Therefore, this paper chose to use a soft thresholding function to process parameter thresholding. The formula for the soft threshold function is as follows:

$$w_{thr} = \begin{cases} [\mathrm{sgn}(w)](|w| - \lambda), & |w| \geq \lambda \\ 0, & |w| < \lambda \end{cases} \tag{6}$$

where w is the wavelet coefficient and λ indicates the threshold calculated according to the threshold principle.

3.3. Bidirectional Long Short-Term Memory Network

Long short-term memory (LSTM) network is a variation of recurrent neural network (RNN) proposed to mitigate two problems: the potential loss of information when processing lengthy sequences, and the gradient can become quite low because of the high number of mathematical operations performed during the processing while remaining far from reaching the threshold [30]. LSTM adds cell states and a series of gate structures on the basis of RNN. It propagates information through forward and backward methods. Bidirectional long short-term memory (BiLSTM) considers fully the interrelationship between time series data, which enables it to extract the time characteristics of signals more effectively.

The LSTM layer is linked by a number of memory units. The interior of each unit can be divided into the forgetting gate, the input gate, and the output gate according to the function. The internal structure of the unit is shown in Figure 3 (Please refer to the following text for explanations of relevant symbols and arrows).

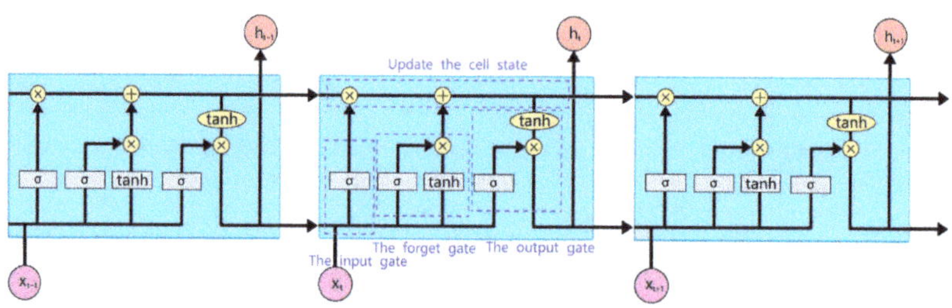

Figure 3. The internal structure of LSTM.

The forget gate (denoted as f_t) determines whether to retain or discard the status information of the previous unit and filters important information into the next unit. The relevant calculation formula is as follows:

$$f_t = \sigma\left(W_f[h_{t-1}, x_t] + b_f\right) \tag{7}$$

where W_f and b_f are the forgetting gate weight and bias items, σ is the sigmoid activation function, and $[h_{t-1}, x_t]$ is the matrix composed of the output information h_{t-1} of the previous moment and the input information x_t of the current moment.

The input gate (denoted as i_t) is used to update the information of the short-term memory unit, and remember the important information of the previous unit and the current unit status. The relevant calculation formula is as follows:

$$i_t = \sigma(W_i[h_{t-1}, x_t] + b_i) \tag{8}$$

$$\widetilde{C}_t = \tanh(W_C[h_{t-1}, x_t] + b_C) \tag{9}$$

where i_t is the input gate control unit, W_i is the input layer weight, and b_i is the input gate bias value; a $\widetilde{C}_t$ is the candidate state created by the tanh layer, W_C represents the weight in the tanh function, and b_C represents the bias in the tanh function.

After passing through the forgetting gate and the input gate, the cell state is updated, and the relevant calculation formula is as follows:

$$C_t = f_t C_{t-1} + i_t \widetilde{C}_t \tag{10}$$

where C_{t-1} is the state of the cell at the previous time, and C_t is the state of the cell at the current time.

The output gate (denoted as o_t) determines the output information after the status update. The relevant calculation formula is as follows:

$$o_t = \sigma(W_o[h_{t-1}, x_t] + b_o) \tag{11}$$

$$h_t = o_t \tanh(C_t) \tag{12}$$

where W_o is the weight of the output and b_o is the offset value of the output; h_t is the output vector at the current time.

On the basis of LSTM, BiLSTM connects forward and backward hidden states to form a new hidden state, realizing the fusion of past and future information. The expression of the BiLSTM network is shown in the following form:

$$H_t = \overrightarrow{h_t} + \overleftarrow{h_t} \tag{13}$$

where $\overrightarrow{h_t}$ is the output of the last forward LSTM layer; $\overleftarrow{h_t}$ is the output of the last backward LSTM layer; and H_t is the final output of the BiLSTM network.

3.4. Sparrow Search Algorithm

Sparrow search algorithm (SSA) is a new optimization algorithm proposed by Xue et al. [31] (2020). It simulates the foraging process of sparrows to obtain solutions to the optimization problem. There are producers and scroungers among sparrows. In addition, there is also the existence of those who are aware of the danger. Producers are responsible for finding food and have higher energy reserves. Scroungers follow and monitor producers and have low energy reserves; some scroungers compete with producers for food. When predators appear in a foraging area, a sparrow recognizes the danger and immediately enters alert mode [32]. The initial positions of these sparrows that are aware of the danger are randomly generated in the population, accounting for 10% to 20%.

In order to survive better, when in a safe range, the producer enters the wide search mode. If it exceeds the safety range, it means that some sparrows have discovered the predator, and all sparrows need to quickly fly to other safe areas. The position update formula for the producer is the following form:

$$X_{i,j}^{t+1} = \begin{cases} X_{i,j}^t \cdot \exp\left(\frac{-i}{\alpha \cdot iter_{\max}}\right) & \text{if } R_2 < ST \\ X_{i,j}^t + Q \cdot L & \text{if } R_2 \geq ST \end{cases} \tag{14}$$

where t indicates the current iteration, $j = 1, 2, \ldots, d$; $X_{i,j}^t$ represents the value of the jth dimension of the ith sparrow at iteration t; $iter_{\max}$ is a constant with the largest number of iterations; $\alpha \in (0, 1]$ is a random number; R_2 ($R_2 \in [0, 1]$) and ST ($ST \in [0.5, 1.0]$) represent the alarm value and the safety threshold, respectively; Q is a random number which obeys normal distribution; and L shows a matrix of $1 \times d$ for which each element inside is 1.

The position update formula for the scrounger is described as follows:

$$X_{i,j}^{t+1} = \begin{cases} Q \cdot \exp\left(\frac{X_{worst}^t - X_{i,j}^t}{i^2}\right) & \text{if } i > \frac{n}{2} \\ X_P^{t+1} + \left|X_{i,j}^t - X_P^t\right| \cdot A^+ \cdot L & \text{otherwise} \end{cases} \tag{15}$$

where X_P is the optimal position occupied by the producer, X_{worst} denotes the current global worst location. A represents a matrix of $1 \times d$ for which each element inside is randomly assigned 1 or -1, and $A^+ = A^T(AA^T)^{-1}$. When $i > \frac{n}{2}$, it suggests that the ith scrounger with the worse fitness value is most likely to be starving.

The mathematical model can be expressed as follows:

$$X_{i,j}^{t+1} = \begin{cases} X_{best}^t + \beta \cdot \left| X_{i,j}^t - X_{best}^t \right| & \text{if } f_i > f_g \\ X_{i,j}^t + K \cdot \left(\frac{\left| X_{i,j}^t - X_{worst}^t \right|}{(f_i - f_w) + \varepsilon} \right) & \text{if } f_i = f_g \end{cases} \tag{16}$$

where X_{best} is the current global optimal location, β is the step size control parameter. $K \in [-1, 1]$ is a random number; Here f_i is the fitness value of the present sparrow; f_g and f_w are the current global best and worst fitness values; and ε is the smallest constant.

4. Data Processing

The original production data from the oilfield usually contains interference information, such as missing values and duplicate values, which will reduce the accuracy of fault prediction. Before model training, it is necessary to conduct data preprocessing on the obtained data with the aim of highlighting the effective information related to different feature data. The pre-processed data was divided into training set, validation set, and testing set according to a certain proportion, and then put into model training, model validation, and model testing, respectively.

4.1. Data Preprocessing

(1) Eliminating unique values

In general, the presence of unique data attributes does not significantly impact the distribution rule and feature correlation within the sample. However, during LSTM model training, constant data for certain features can still influence the final results. To mitigate this issue, this study removed such data items.

(2) Dealing with missing values

Missing values can be categorized into three types based on their distribution: completely random, random, and non-random. The presence of missing values directly leads to a loss of valuable information in the system and incomplete variables can disrupt or even render the training process impossible, thereby reducing the proportion of deterministic components displayed by the system and increasing errors. Typically, samples containing missing values are either deleted outrightly or filled using interpolation methods. Given that there are only a small number of missing values in our dataset of pumping unit wells, we have chosen to exclude wells with missing data from our training set. Additionally, excessively small or large data should be removed to minimize errors.

(3) Feature encoding

When learned features contain non-numeric symbols, they need to be encoded for quantification purposes. One-hot encoding is often used when dealing with few eigenvalues to visually distinguish non-numeric features. One-hot coding is also known as one-bit effective coding. It uses N-bit registers to encode N states, and uses 1 and 0 to indicate the true or false of the corresponding state. Only one is valid at any time. In this paper, the five major reasons for the failure of the pumping unit are obtained statistically: wear (00001), fatigue (00010), scale (00100), wax (01000), and corrosion (10000).

(4) Data denoising

Preliminary model testing has shown that directly training processed data—after removing unique values, handling missing values, and standardization—results in generally low prediction accuracy at around 50%. In order to improve model prediction accuracy, further denoising of our obtained dataset is necessary. Wavelet threshold denoising technology and variational mode decomposition technology have demonstrated promising results in signal processing applications. This paper will combine the strengths of both parties to achieve the goal.

This paper selected heursure threshold principle and soft thresholding function. Based on the data characteristics of each fault, coif4, bior6.8, db3, sym8, and db4 were used as wavelet basis functions. The statistical data contained a total of 14 feature quantities, with a maximum of 90 data retained in each sample. Figure 4 shows the comparison curve of one feature data of the same sample before and after wavelet threshold denoising. The curve after wavelet threshold denoising is significantly smoother than the curve of the unprocessed data, characterizing the non-stationary data well.

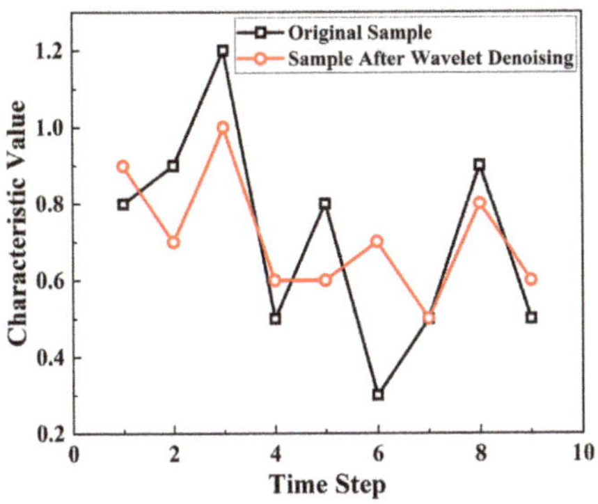

Figure 4. Comparison of characteristic curves before and after wavelet denoising.

By using wavelet basis functions to decompose the original data, features with short time steps could not be decomposed and would be skipped. The selection of time steps resulted in different sample sizes in the obtained dataset. This paper chose a threshold of 90 to filter out data with many missing values or too short time steps, leaving 714 samples.

VMD can smoothen the image while preserving the detailed components of the original signal well. After multiple experiments, it was found that K was set to 14, α was set to 250, *tol* was set to 5×10^{-6}, and *tau* was set to 0.01. At this point, the effect of VMD reached its best. Figure 5 shows a comparison image of data with a time step of 800 before and after VMD. From the decomposition diagram, it can be seen that VMD converts the initial signal into non-recursive signals IMF1~IMF14, which represent the trend at different time scales and frequency ranges. It is the variation trend at different scales that is helpful to judge the fault type.

(5) Data standardization

When statistical data was collected, there were differences between units. If data with different units was used directly, data with larger numerical values would dominate, reducing iteration speed and affecting prediction results. To make the training process stable and efficient, this paper adopted the standard score method to uniformly process the data.

(6) Dataset partitioning

After data preprocessing and filtering, a total of 714 sets of data were obtained, with each array representing a well. The dataset was divided into training set, validation set, and testing set: 426 groups were the training set (accounting for 60% of the total dataset), 144 groups were the validation set (accounting for 20% of the total dataset), and 144 groups were the testing set (accounting for 20% of the total dataset).

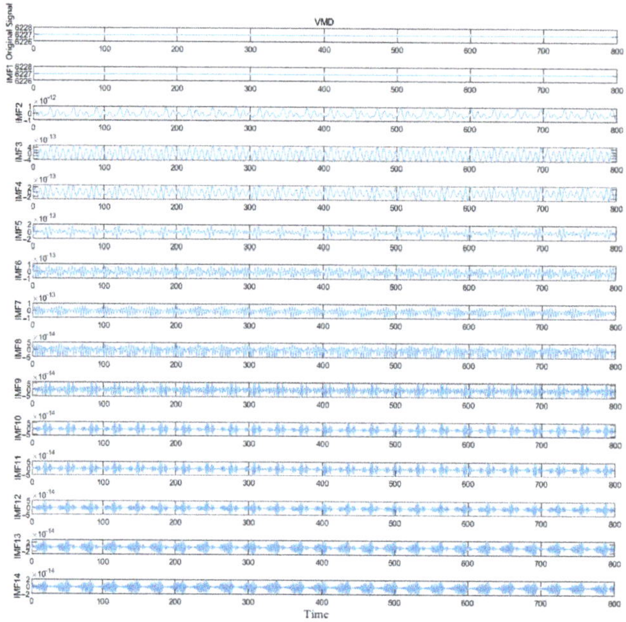

Figure 5. VMD results.

4.2. Model Training and Validation

Appropriate parameters are particularly important for the training results of BiLSTM models. Parameter settings need to balance the complexity and generalization ability of the model, otherwise it may cause overfitting or underfitting problems. Therefore, this paper introduced SSA to simulate the foraging and predatory behavior of sparrows and retrieve the optimal parameter combination for the BiLSTM model.

After several optimizations, the optimal parameter combination was obtained as follows: the optimal number of hidden units was 165, the optimal maximum training period was 88, and the optimal initial learning rate was 0.0066.

This paper used an adaptive moment estimation (ADAM) optimizer to train on the training set, with a total of 88 rounds of training, four iterations per round, and a total of 352 iterations. To avoid gradient explosion, the gradient threshold was set to 1. The initial learning rate was 0.0066, and the learning strategy was adjusted in stages. When to reduce the learning rate was determined by the learning rate decline cycle. In this article, it was set to 20, that is, every 80 iterations will reduce one cycle. The BiLSTM network was trained according to the above model parameters. The specific parameters are shown in Table 2.

Table 2. Structure of the Model.

Parameter	Description
Problem	fault prediction of sucker rod pumping wells
Algorithm	VMD-SSA-BiLSTM
Input	The production journal of oil wells
Output	Fault Type
Data Division	Training (60%), Validation (20%), Testing (20%)
Hidden Layer Size	165 neurons

4.3. Model Testing

As this article focuses on fault type diagnosis, there is only a distinction between 0 and 1 error, which leads to the loss of reference value for indicators such as RMSE, MSE, and MSE. Therefore, this paper chose accuracy as the sole evaluation indicator.

The test set was put into the trained model to test its effectiveness. Figure 6 shows the comparison curve between the model fault prediction results and the actual results. To render the results more effectively, this paper categorized and ranked the test well samples based on their fault types before generating the visualization. It could be seen from the figure that the two curves highly overlap. Among the 144 selected test wells, 142 wells had the same test results as the actual results, with a prediction accuracy of 98.61%, indicating good testing results.

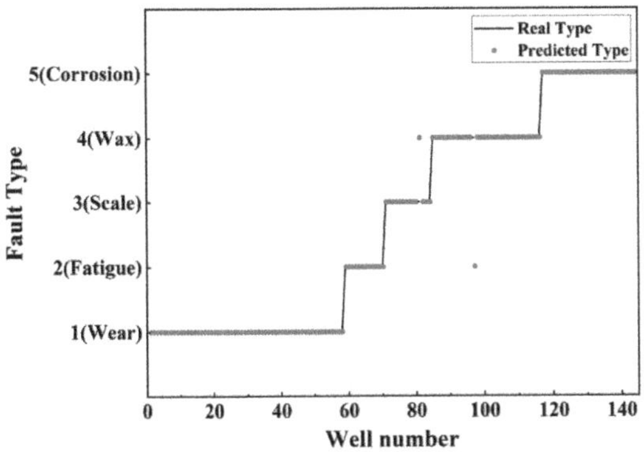

Figure 6. Comparison between the predicted type and the real type.

5. Model Assessment

To demonstrate the effectiveness of the model, different models were tested under the same environment and parameters, and the accuracy of each model was shown in Table 3. As indicated by Table 3, there was a certain improvement in the accuracy of each model after using VMD or optimizing the training model using SSA.

Table 3. The effect of different models in the same environment.

Model	Total Number of Wells	Number of Correct Predictions	Accuracy Rate
BiLSTM	144	101	70.13%
LSTM	144	104	72.22%
GRU	144	84	58.33%
VMD-BiLSTM	144	141	97.92%
VMD-LSTM	144	139	96.53%
VMD-GRU	144	137	95.14%
SSA-BiLSTM	144	105	72.91%
SSA-LSTM	144	105	72.91%
SSA-GRU	144	88	61.11%
VMD-SSA-BiLSTM	144	142	98.61%
VMD-SSA-LSTM	144	139	96.53%
VMD-SSA-GRU	144	138	95.83%

The predicted results of the VMD-SSA-BiLSTM model, VMD-SSA-LSTM model, and VMD-SSA-GRU model were compared with the actual values, respectively. Among them, BiLSTM performed best. The results are shown in Figure 7.

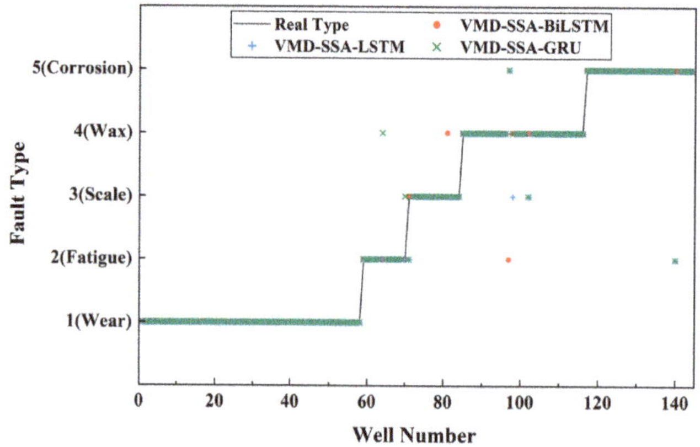

Figure 7. Accuracy comparison of different models.

The experimental results showed that the prediction accuracy of the model can be improved by inputting decomposed data into the model, which proved that VMD can reduce the noise and complexity of the original data. SSA could help model parameter selection and improve prediction accuracy. The prediction accuracy of VMD-SSA-BiLSTM is up to 98.61%, which has broad application prospects.

In addition, for reasons of analyzing the influence of the randomness of the test set on the prediction accuracy, the prediction score is used to further evaluate the prediction effect of the model. The value of the prediction score represents the probability that a fault that will occur in a certain well is predicted to be this type of fault, and the sum of prediction scores of the five types of faults in each test well is 1. As shown in Figure 8, among the 144 test wells, the prediction scores of a class of faults in most test wells are close to 1, which indicates that the model achieves great generalization capability and robustness.

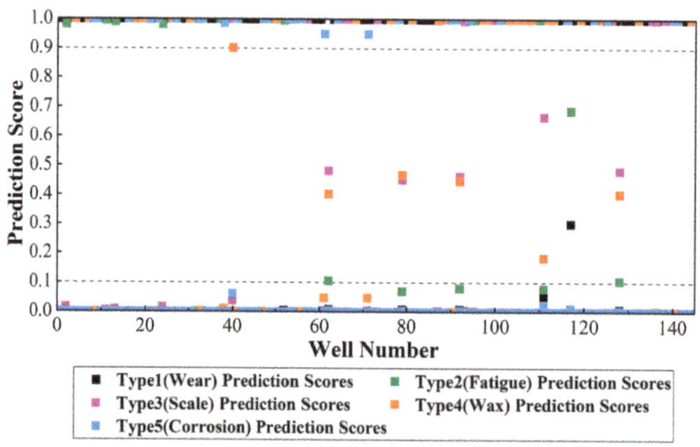

Figure 8. Prediction scores.

6. Conclusions

This paper aims to improve resource utilization and save energy by studying the fault warning of the rod pumping system. Our study contributes to oilfield development by proposing a hybrid model based on artificial intelligence that can provide fault warning, while considering the multiple non-linear interactions between components involved in such a complex production system. The large amount of data used in this paper was obtained through cooperation with Xinjiang Oilfield, fully considering practicality. In addition, this paper bridges the gap that very little research has been published on using production journals to improve oilfield production safety conditions.

In this study, deep learning models, especially BiLSTM, VMD, and SSA, have proven to be effective in helping predict faults in the rod pumping system. The outstanding accuracy of the model proposed in this article in the actual production activities of Xinjiang Oilfield, reaching 98.61%, shows that using this type of model is conducive to improving the safety level of oilfields and reducing production risks. In the current situation where the effectiveness of oilfield data is not high, this hybrid model still has good robustness. But at the same time, we suggest that oilfields emphasize the importance of maintaining database consistency to improve the predictive ability of deep learning models and enhance the safety level of oilfields.

In future research, more input variables can be considered, especially parameters that can reflect the degree of human operational norms. This would reduce safety incidents caused by human factors and provide stronger security for oilfield safety. Furthermore, with the advancement of smart oilfields, the model presented in this paper can be expanded to include more inputs and larger datasets. The emergence of more effective real-time data will provide more insightful results.

Author Contributions: Conceptualization, A.Z. and Y.Z.; methodology, A.Z.; investigation, X.L., X.F., X.R. and L.Y.; resources, Q.L.; writing—original draft preparation, A.Z.; writing—review and editing, A.Z.; supervision, Y.Z.; All authors have read and agreed to the published version of the manuscript.

Funding: This research was funded by National Natural Science Foundation of China [grant number 52004301] and "Western young scholars" of Chinese Academy of Sciences [grant number 2021-XBQNXZ-033].

Data Availability Statement: Restrictions apply to the availability of these data. Data were obtained from Xianjiang Oilfield and are available from the authors with the permission of Xinjiang Oilfield.

Acknowledgments: The authors appreciate the support of the National Natural Science Foundation of China and the Chinese Academy of Sciences. Authors also would like to extend their appreciation to PetroChina Xinjiang Oilfield Company for providing data support.

Conflicts of Interest: Author Qingxia Li was employed by the company PetroChina Xinjiang Oilfield Company. The remaining authors declare that the research was conducted in the absence of any commercial or financial relationships that could be construed as a potential conflict of interest.

References

1. Kumar, A.; Upadhyay, R.; Kumar, S. Tubing and Rod Failure Analysis in Rod Pumped Wells in an Indian Western Oil Field. *SPE J.* **2023**, *28*, 1481–1501. [CrossRef]
2. Mohammed, A. Data driven-based model for predicting pump failures in the oil and gas industry. *Eng. Fail. Anal.* **2023**, *145*, 107019. [CrossRef]
3. Cao, L.; Yuan, H.; Pan, Z.; Liu, Z.; Zhang, B.; Sun, T.; Liu, J.; Wu, H. Dynamic Scaling Prediction Model and Application in Near-Wellbore Formation of Ultradeep Natural Gas Reservoirs. *SPE J.* **2024**, *29*, 2476–2493. [CrossRef]
4. Peng, X.; Yu, H.; Lian, Z.; Liu, T.; Shi, J.; Yang, D. Study on Fatigue Behavior of S135 Steel and Titanium Alloy Drillpipes: Experiment and Modeling. *SPE Drill. Complet.* **2023**, *38*, 261–272. [CrossRef]
5. Company, R.R.O.; Timashev, E.; Urazakov, K. Optimization of the operating mode of sucker-rod pump units based on the quantitative analysis of dynamometer cards (Russian). *Oil Ind. J.* **2022**, *2022*, 38–41. [CrossRef]
6. Liu, C.; Zhang, K.; Deng, Z.; Zhao, X.; Zhang, X.; Wang, Z. A Failure Risk Assessment Method for Lithium-ion Batteries Based on Big Data of After-sales Vehicles. *Eng. Fail. Anal.* **2024**, *163*, 108559. [CrossRef]
7. Lenwoue, A.R.K.; Li, Z.; Tang, C.; Zhang, W.; Ding, S.; Hu, P.; Sun, W. Recent Advances and Challenges of the Application of Artificial Intelligence to Predict Wellbore Instabilities during Drilling Operations. *SPE Drill. Complet.* **2023**, *38*, 645–662. [CrossRef]
8. Gilbert, W.E. *An Oil-Well Pump Dynagraph*; American Petroleum Institute: Washington, DC, USA, 1936.
9. Gibbs, S.; Neely, A. Computer Diagnosis of Down-Hole Conditions in Sucker Rod Pumping Wells. *J. Pet. Technol.* **1966**, *18*, 91–98. [CrossRef]
10. Derek, H.; Jennings, J.; Morgan, S. Sucker rod pumping unit diagnostics using an expert system. In Proceedings of the Permian Basin Oil and Gas Recovery Conference, Midland, TX, USA, 10–11 March 1988; OnePetro: Richardson, TX, USA, 1988. [CrossRef]
11. Foley, W.; Svinos, J. Expert adviser program for rod pumping (includes associated paper 19367). *J. Pet. Technol.* **1989**, *41*, 394–400. [CrossRef]

12. Schirmer, P.; Gay, J.C.; Toutain, P. Use of advanced pattern-recognition and knowledge-based system in analyzing dynamometer cards. *SPE Comput. Appl.* **1991**, *3*, 21–24. [CrossRef]
13. Rogers, J.D.; Guffey, C.G.; Oldham, W.J.B. Artificial neural networks for identification of beam pump dynamometer load cards. In Proceedings of the SPE Annual Technical Conference and Exhibition, New Orleans, LA, USA, 23–26 September 1990; OnePetro: Richardson, TX, USA, 1990. [CrossRef]
14. Nazi, G.; Ashenayi, K.; Lea, J.; Kemp, F. Application of Artificial Neural Network to Pump Card Diagnosis. *SPE Comput. Appl.* **1994**, *6*, 9–14. [CrossRef]
15. Gamal, H.; Elkatatny, S.; Al Gharbi, S. Rig Sensor Data for AI-ML Technology-Based Solutions: Research, Development, and Innovations. In Proceedings of the ADIPEC, Abu Dhabi, United Arab Emirates, 2–5 October 2023. [CrossRef]
16. Gao, Q.; Sun, S.; Liu, J. Working Condition Detection of Suck Rod Pumping System via Extreme Learning Machine. In Proceedings of the 2015 IEEE International Conference on CYBER Technology in Automation, Control, and Intelligent Systems (CYBER), London, UK, 13–14 March 2015; pp. 503–507. [CrossRef]
17. Lv, X.; Wang, H.; Zhang, X.; Liu, Y.; Jiang, D.; Wei, B. An Evolutional SVM Method Based on Incremental Algorithm and Simulated Indicator Diagrams for Fault Diagnosis in Sucker Rod Pumping Systems. *J. Pet. Sci. Eng.* **2021**, *203*, 1–17. [CrossRef]
18. Yin, C.; Zhang, K.; Zhang, L.; Wang, Z.; Liu, P.; Zhang, H.; Yang, Y.; Yao, J. Imbalanced Working States Recognition of Sucker Rod Well Dynamometer Cards Based on Data Generation and Diversity Augmentation. *SPE J.* **2023**, *28*, 1925–1944. [CrossRef]
19. He, Y.-P.; Cheng, H.-B.; Zeng, P.; Zang, C.-Z.; Dong, Q.-W.; Wan, G.-X.; Dong, X.-T. Working condition recognition of sucker rod pumping system based on 4-segment time-frequency signature matrix and deep learning. *Pet. Sci.* **2024**, *21*, 641–653. [CrossRef]
20. Zheng, B.; Gao, X.; Li, X. Fault detection for sucker rod pump based on motor power. *Control. Eng. Pract.* **2019**, *86*, 37–47. [CrossRef]
21. Wei, J.; Gao, X. Fault diagnosis of sucker rod pump based on deep-broad learning using motor data. *IEEE Access* **2020**, *8*, 222562–222571. [CrossRef]
22. Bai, T.; Li, X.; Ding, S. Research on Electrical Parameter Fault Diagnosis Method of Oil Well Based on TSC-DCGAN Deep Learning. In Proceedings of the 2022 3rd International Conference on Big Data, Artificial Intelligence and Internet of Things Engineering (ICBAIE), Xi'an Shiyou University, Xi'an, China, 15–17 July 2022; pp. 753–761. [CrossRef]
23. Liu, C.D.; Yang, Y.P.; Li, L.Y.; You, Y.W. Diagnosis of Loose Transformer Winding Faults in Pumped Storage Power Station Based on NGO-VMD and LSSVM. In Proceedings of the 2024 9th Asia Conference on Power and Electrical Engineering (ACPEE), Shanghai, China, 11–13 April 2024; pp. 2494–2498. [CrossRef]
24. Dragomiretskiy, K.; Zosso, D. Variational mode decomposition. *IEEE Trans. Signal Process.* **2014**, *62*, 531–544. [CrossRef]
25. Parri, S.; Teeparthi, K.; Kosana, V. A hybrid methodology using VMD and disentangled features for wind speed forecasting. *Energy* **2024**, *288*, 129824. [CrossRef]
26. de Loynes, B.; Navarro, F.; Olivier, B. Data-driven thresholding in denoising with Spectral Graph Wavelet Transform. *J. Comput. Appl. Math.* **2021**, *389*, 113319. [CrossRef]
27. Song, F.; Ding, H.; Wang, Y.; Zhang, S.; Yu, J. A Well Production Prediction Method of Tight Reservoirs Based on a Hybrid Neural Network. *Energies* **2023**, *16*, 2904. [CrossRef]
28. Sun, Y.; Cao, Y.; Li, P.; Su, S. Sound Based Degradation Status Recognition for Railway Point Machines Based on Soft-Threshold Wavelet Denoising, WPD, and ReliefF. *IEEE Trans. Instrum. Meas.* **2024**, *73*, 1–9. [CrossRef]
29. Du, E.; Liu, Y.; Cheng, Z.; Xue, L.; Ma, J.; He, X. Production Forecasting with the Interwell Interference by Integrating Graph Convolutional and Long Short-Term Memory Neural Network. *SPE Reserv. Eval. Eng.* **2022**, *25*, 197–213. [CrossRef]
30. Carpenter, C. AI-Based System Automates Textual Classification of Daily Drilling Reports. *J. Pet. Technol.* **2024**, *76*, 55–57. [CrossRef]
31. Xue, J.; Shen, B. A novel swarm intelligence optimization approach: Sparrow search algorithm. *Syst. Sci. Control. Eng.* **2020**, *8*, 22–34. [CrossRef]
32. Wu, C.; Fu, X.; Pei, J.; Dong, Z. A Novel Sparrow Search Algorithm for the Traveling Salesman Problem. *IEEE Access* **2021**, *9*, 153456–153471. [CrossRef]

Disclaimer/Publisher's Note: The statements, opinions and data contained in all publications are solely those of the individual author(s) and contributor(s) and not of MDPI and/or the editor(s). MDPI and/or the editor(s) disclaim responsibility for any injury to people or property resulting from any ideas, methods, instructions or products referred to in the content.

energies

MDPI

Article

Research on the Liquid Helium Insulation Characteristics of an Experimental System

Ye Chen [1,2], Liang Guo [1,2], Qiming Jia [1], Xiujuan Xie [1,*], Weiping Zhu [1] and Ping Wang [1]

[1] State Key Laboratory of Cryogenic Science and Technology, Technical Institute of Physics and Chemistry, Chinese Academy of Sciences, Beijing 100190, China
[2] University of Chinese Academy of Sciences, Beijing 100049, China
* Correspondence: xiexiujuan@mail.ipc.ac.cn

Abstract: The research on the thermal insulation performance of experimental systems in the liquid helium temperature range is relatively scarce. This paper presents the theoretical design and establishment of a liquid helium storage system for insulation research, consisting of a liquid helium Dewar, a daily boil-off rate test subsystem, and a helium recovery subsystem. The passive thermal insulation structure consisted of a multilayer insulation (MLI) system with hollow glass microspheres serving as spacers. Based on self-built data acquisition, experiments were conducted to investigate the liquid helium insulation characteristics of an experimental system. A theoretical thermal analysis of the Dewar was conducted, resulting in the derivation of an expression for the heat leak of the Dewar. The analysis indicates that the evaporation capacity from the liquid helium Dewar was significantly affected by the structure of the neck tube. The overall relative error between the simulated and experimental temperature distribution of the insulation layer is 14.3%, with a maximum error of 22.3%. The system had an average daily boil-off rate of 14.4%, a heat leakage of 7.5 W, and a heat flux of 2.254 W/m^2, while the effective thermal conductivity of the MLI with hollow glass microspheres was determined to be 2.887×10^{-4} W/(m·K). Furthermore, the apparent thermal conductivity between different layers of MLI significantly fluctuated with increasing temperature, ranging from a maximum of 5.342×10^{-4} W/(m·K) to a minimum of 1.721×10^{-4} W/(m·K).

Keywords: liquid helium; thermal insulation characteristics; hollow glass microspheres; daily boil-off rate; multilayer insulation

1. Introduction

The unique fluid properties of liquid helium, such as its extremely low temperature, low latent heat of vaporization, and high thermal conductivity, make it one of the most commonly utilized cryogens [1]. It plays an irreplaceable role in various fields, including cryogenic superconductivity, high-energy particle physics, cryogenic medicine, and quantum computing [2].

Liquid helium plays a crucial role in numerous scientific fields; therefore, enhancing the thermal insulation performance for the storage of liquid helium is of significant importance. Compared with other cryogenic fluids such as liquid natural gas (110 K), liquid nitrogen (77 K), and liquid hydrogen (20 K), there are more challenges with achieving liquid helium storage with the use of thermal insulation materials at 4.2 K. Thermal insulation structures typically involve both passive and active heat transfer mechanisms. Passive thermal insulation structures that are currently applicable in the temperature range of liquid helium include spray-on foam insulation (SOFI), multilayer insulation, and hollow glass microspheres (HGMs). On the other hand, active thermal insulation methods involve active refrigeration at low temperatures to minimize the evaporation of cryogenic fluids, achieving zero-boil-off (ZBO) conditions [3].

Researchers have conducted extensive studies on various thermal insulation materials. Tseng et al. [4] experimentally measured the thermal conductivity of SOFI in the temperature range of 20 K, achieving values as low as 5×10^{-3} W/(m·K). Liu et al. [5] simulated an MLI system and reported that its thermal conductivity ranged from 10^{-6} to 10^{-5} W/(m·K) in liquid hydrogen environments, noting the susceptibility of the MLI to high vacuum and its high cost. The thermal conductivities of HGMs, known for their low density, high strength, and low thermal conductivity, were measured by M.S. Allen et al. [6] under a vacuum of 1×10^{-3} Pa. The HGMs achieved a thermal conductivity as low as 7×10^{-4} W/(m·K) in the temperature range of 77–293 K.

Multi-Layer Insulation (MLI) exhibits exceptional thermal insulation properties. Naes et al. [7] conducted tests on the thermal performance of MLI within the temperature range of 4.2 K to 100 K, yielding a heat flux of 0.75 W/m^2. M. Vanderlaan et al. [8] performed repeated tests on 25 layers of MLI at 20 K, observing a heat flux variation ranging from 0.98 to 1.15 W. Additionally, Q.S. Shu et al. [9] indicated that the optimal number of layers for MLI at 77 K is between 30 and 40. P.J. Sun et al. [10] examined the thermal insulation performance of MLI under varying pressure conditions, revealing that as the thermal boundary temperature increases, the heat transfer coefficient also rises, with this effect becoming more pronounced as the vacuum quality deteriorates. Based on experimental findings, S.L. Bapat et al. [11] proposed that during the heat transfer process in MLI, the gap pressure must be significantly higher than the pressure within the vacuum chamber, thereby increasing the contribution of gas conduction.

Furthermore, extensive research has been conducted on composite thermal insulation materials. Hastings et al. [12] performed experimental studies on SOFI and variable density multilayer insulation (VD-MLI) at liquid nitrogen temperatures and reported a heat leakage of 0.31 W/m^2. Hedayat et al. [13] tested composite thermal insulation materials, including SOFI and MLI, on the MHTB platform, showing a heat flux as low as 0.22 W/m^2 at liquid hydrogen temperatures. Zheng et al. [14] quantitatively analyzed optimization strategies for MLI + VCS (vapor-cooled shield) and VD-MLI + VCS composite thermal insulation structures via a thermodynamic model, identifying potential reductions in the heat flux of 58.05% and 66.32%, respectively, at liquid hydrogen temperatures. Zhang et al. [15] developed thermodynamic models to analyze various composite thermal insulation structures within the liquid helium temperature range and determined that the optimal positions for VCS were 30% from the cold boundary for SOFI + MLI + VCS and 25% for SOFI + VD-MLI + VCS. Zheng et al. [16] discovered that when subjected to a vacuum of 10^{-3} Pa and a cold boundary temperature of 20 K, the incorporation of HGM + VCS resulted in a remarkable reduction in the heat flux of 58.08% compared with that of HGMs.

To summarize, thermal insulation materials have evolved, from low thermal conductivity solid materials to high-vacuum multilayer insulation (MLI) materials, and eventually to composite thermal insulation structures. These materials are applied in various use cases with differing thermal insulation requirements. For instance, in the 77 K temperature range, SOFI or HGMs are commonly used, while in the 20 K range, where higher thermal insulation performance is demanded, MLI and composite insulation structures are more frequently employed. However, for the 4 K temperature range, which requires extremely high insulation performance, the scarcity of liquid helium and its tendency to evaporate rapidly complicate the situation. As a result, there is a lack of sufficient data on the insulation performance of structures for this temperature range, making it difficult to optimize designs for such applications. MLI is regarded as the most effective thermal insulation method. Hollow glass microspheres, in comparison to other spacer materials, offer advantages such as low thermal conductivity, high strength, and resistance to deformation. Therefore, in this study, an experimental platform was established to investigate the insulation characteristics of MLI in the liquid helium temperature range through theoretical design and experimental research. This platform utilizes MLI with hollow glass microspheres (HGMs) as spacers [17–19], which are chosen for their high strength and low thermal conductivity, as the passive thermal insulation structure. Liquid helium is employed as the working fluid in our experiments to evaluate and analyze the insulation performance of the experimental system.

2. Theoretical Design of Experimental System for Liquid Helium Insulation Characteristics

2.1. PFD of Experimental System for Liquid Helium Insulation Characteristics

The test system for evaluating liquid helium insulation is illustrated in the PFD in Figure 1. It consisted of four primary components: a liquid helium Dewar, a liquid helium daily boil-off rate test subsystem, and a helium recovery subsystem. These components were interconnected via bellows tubes, and their functionality was regulated by check valves to enable various operational modes. This experimental setup facilitated comprehensive evaluations of both the daily boil-off rates characteristics of liquid helium.

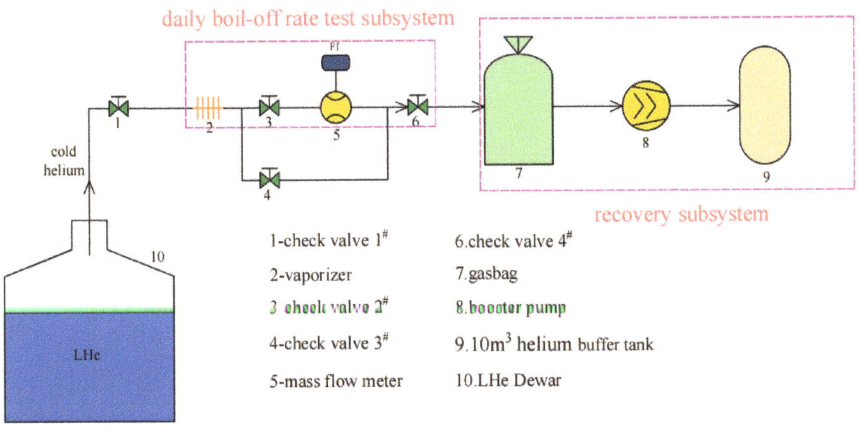

Figure 1. PFD of the experimental liquid helium storage system.

2.2. Theoretical Analysis of Heat Leakage from a Liquid Helium Dewar

The heat leakage in each subsystem mentioned above was primarily attributed to the liquid helium Dewar. A schematic cross-sectional diagram of the liquid helium Dewar is shown in Figure 2, which illustrates the insulated supports between the inner and outer tanks, namely, the inner tank, outer tank, and neck tube. The external surface of the liquid helium reservoir was covered in MLI for thermal insulation purposes.

Cernox-type sensors T01–T03 with an accuracy of ±4 mK @ 4.2 K were uniformly distributed along the entire outer surface of the neck tube from top to bottom. Cernox-type sensors (Lakeshore, Columbus, OH, USA) T11–T13 and PT100 type T14–T16 with an accuracy error of ±10 mK @ 77 K, were evenly distributed along the insulation layers, specifically at layers 0, 12, 24, 36, 48, and 60.

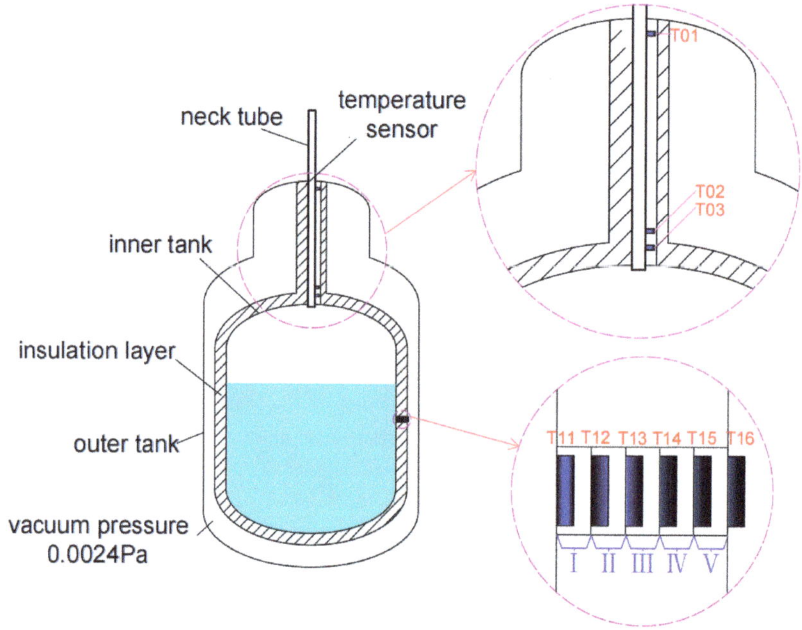

Figure 2. Schematic cross-sectional diagram of the liquid helium Dewar.

Therefore, as illustrated in Figure 3 and outlined in previous work [20], heat leakage from the Dewar could be categorized into three parts: heat conduction through the neck tube Q_{nt}, heat transfer across the insulation layer Q_{il}, and heat transfer from helium gas within the neck tube Q_g. Thus, heat leakage from the liquid helium Dewar was quantified via the following equation:

$$Q_{dewar} = Q_{nt} + Q_{il} + Q_g \tag{1}$$

$$Q_{nt} = \frac{\overline{A}_{nt}}{L} \int \lambda_{ntT} dT \tag{2}$$

$$Q_{il} = \frac{2\pi\overline{\lambda}_{il}\overline{l}(T_h - T_c)}{\ln\left(\frac{r_h}{r_c}\right)} \tag{3}$$

$$Q_g = \frac{\overline{A}_g}{L} \int \lambda_{gT} dT \tag{4}$$

The temperature at the bottom measuring point of the neck tube, denoted as T_0(K) and oriented vertically upwards in Figure 3, was found to be 10 K on the basis of previous steady-state experimental observations. Similarly, the temperature at the outlet of the neck tube was measured to be 30 K. Notably, the thermal conductivities of the solid of the neck tubes and the gas within these tubes exhibited minimal sensitivity to pressure variations. To accurately represent this behavior over a range of temperatures from 10 K to 30 K, a polynomial approximation was employed for characterizing the thermal conductivity properties:

$$\lambda_{nt} = 0.1294T - 0.4132 \tag{5}$$

$$\lambda_g = 0.0003T^3 - 0.0263T^2 + 1.5016T + 4.4089 \tag{6}$$

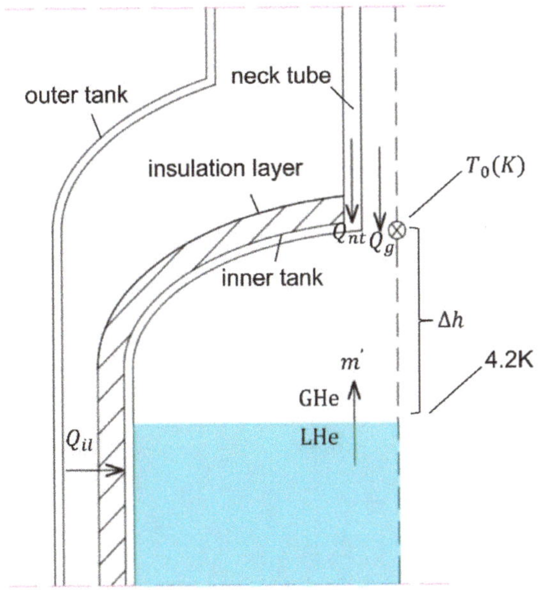

Figure 3. Diagram of showing heat leakage from a liquid helium Dewar.

The apparent thermal conductivity of the insulation layer λ_{ail} was determined through measurements conducted at a physical property testing center, which yielded a value of 7.17×10^{-5} W/(m·K) (77 K–293 K). This parameter was obtained by integrating the given expression. Specifically, $Q_{nt} = 0.016$ W, $Q_{il} = 2.096$ W, and $Q_g = 2.891$ W were utilized in calculating the total heat leakage of $Q_{dewar} = 5.003$ W.

3. Experimental System for Liquid Helium Insulation Characteristics

A picture of the experimental system for liquid helium insulation characteristics is shown in Figure 4.

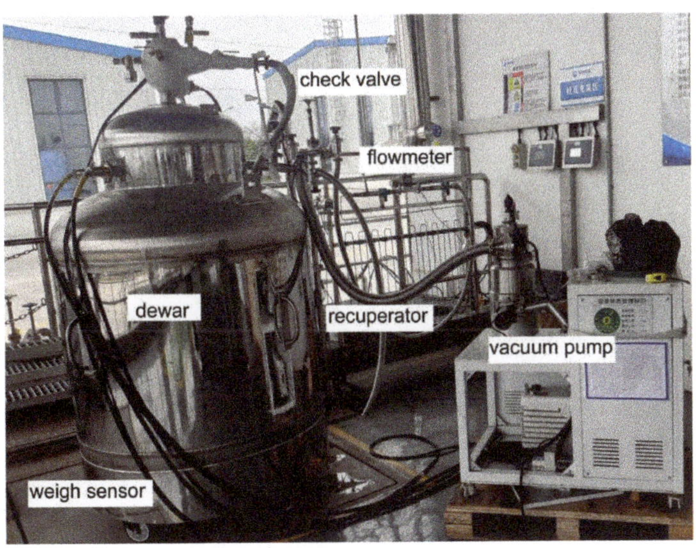

Figure 4. Picture of experimental system for liquid helium insulation characteristics.

The 500 L Dewar (FHF230100, Fullcryo, Zhongshan, China) for liquid helium was vertically structured, featuring inner and outer containers constructed from 304 stainless steel. As illustrated in Figure 2, the inner container has a diameter of 950 mm, featuring a butterfly head structure with a thickness of 3 mm. Notably, the outer layer of the inner container had a high-vacuum multilayer insulation structure. The thermal insulation material consists of a total of 60 layers, utilizing polyester as the reflective layer's film material, featuring a double-sided aluminum-coated film structure. The spacing material consists of glass microspheres, applied to one side of the reflective layer, the diameter of the hollow glass microsphere lattice is 0.2 mm. The arrangement of the spacers and the perforation pattern is illustrated in Figure 5.

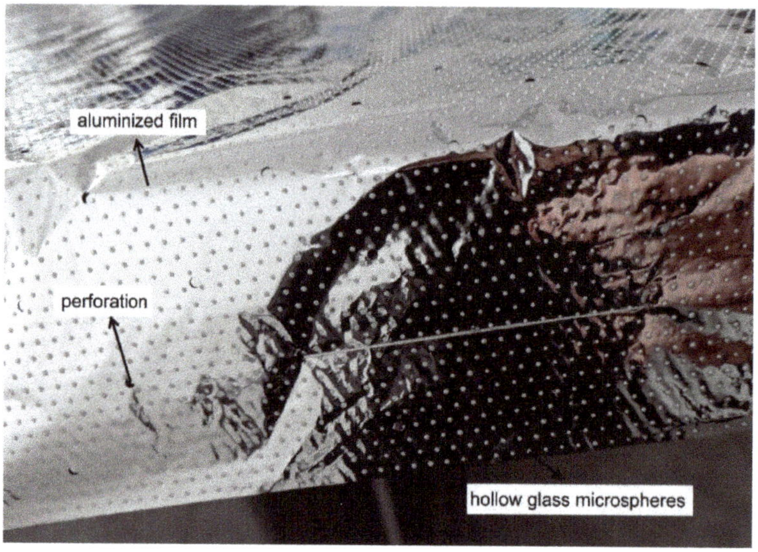

Figure 5. Picture of MLI with hollow glass microspheres.

The subsystem for testing the daily boil-off rate of liquid helium comprised a manual shut-off valve, bellows tube, vaporizer, and ALICAT mass flow meter (Alicat Scientific, Tucson, AZ, USA). The mass flow meter had a range of 0 to 50 SLPM, a response time of $\leq$10 ms, achieved an accuracy better than 1%, and operated within a temperature range of -10 °C–$+$60 °C.

To prevent helium wastage, a helium recovery subsystem was employed, comprising a manual shut-off valve, bellows tube, gasbag, helium recovery compressor, and a 10 m^3 buffer tank. During testing, the helium gas emitted from the system at ambient temperature was directed into the gasbag and subsequently pressurized by the helium recovery compressor before being stored in the high-pressure buffer tank.

As illustrated in Figure 1, liquid helium evaporates within the Dewar to form cold helium gas. During the pre-cooling phase, check valves 1, 3, and 4 are opened. The cold helium gas passes through the vaporizer, warming to ambient temperature, and enters the gas bag. After being pressurized by the booster pump, it is directed into the buffer recovery system. Upon completion of the pre-cooling phase, check valve 2 is closed, and check valve 2 is opened to allow the recovery of gas into the buffer tank. Throughout the experiment, data acquisition equipment is employed to interface with the temperature sensors and flow meters.

The data acquisition interface for the experimental system with liquid helium insulation characteristics is depicted in Figure 6. Temperature sensors T01 to T03, located on the neck tube, and T11 to T16 within the MLI recorded measurements at one-second intervals. After the experimental setup is complete, including vacuuming and leak testing, 90% of the Dewar's 500 L capacity is filled with liquid helium. The Dewar's temperature rapidly decreases, but the internal fluid remains highly unstable due to the large temperature gradient. This stage lasts for 48 h as pre-cooling. After the first resting period, additional liquid helium is added to reach 90% fill again, followed by another 24 h rest period until thermal equilibrium is achieved. This system was subsequently connected to the daily boil-off rate testing subsystem for a continuous test period of 72 h, during which instantaneous readings from the flow meter were recorded once every minute.

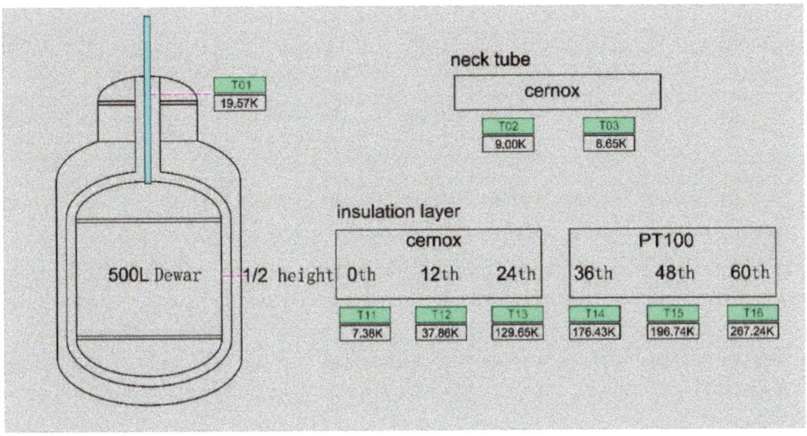

Figure 6. Data acquisition interface of the setup for determining the liquid helium insulation characteristics of the experimental system.

4. Results and Discussion

4.1. Uncertainty Analysis of Experimental System

Experimental errors mainly come from measurement errors and calculation errors. In our experiments, the measurement uncertainties attributable to resistance, the measurement bridge, and temperature control fluctuations are ±2.5 mK, ±1 mK, and ±2.5 mK, respectively. The calibration uncertainties are listed in Table 1 resulting in a combined standard uncertainty for the temperature sensors of ±24.28 mK, equivalent to 1.16%.

Table 1. Calibration uncertainty of temperature sensors.

Temperature \ Model	Cernox	PT100
4.2 K	±4 mK	—
10 K	±4 mK	—
20 K	±8 mK	±9 mK
30 K	±9 mK	±9 mK
50 K	±12 mK	±10 mK
100 K	±16 mK	±11 mK
300 K	±45 mK	±24 mK

The data processing in this experiment focuses on the calculation of heat leakage and apparent thermal conductivity. According to the principles of the cryogenic experiment, the uncertainty analysis for these two parameters primarily encompasses geometry of insulation layer, temperature measurement, and mass flowmeter, as detailed in Table 2.

Table 2. Calibration uncertainty of data processing.

Source of Error	Uncertainty
Geometry of insulation layer	2.05%
Temperature measurement	1.16%
Mass flowmeter	0.5%
Combined standard uncertainty	2.78%

The experimental conditions and stability criteria are shown in Table 3.

Table 3. Experimental conditions for the liquid helium insulation system.

Working Fluid	Precooling Duration	Static Duration	Vacuum Degree	Warm Boundary	Cold Boundary	Stability Criterion
He	72 h	72 h	2.4×10^{-3} Pa	Ambient temperature	4.2 K	delta T < 0.02 degrees C/min

4.2. Analysis of the Daily Boil-Off Rate of the Liquid Helium Dewar

Figure 7 depicts the trend in environmental temperature, with the dashed line indicating the 12 a.m. time point during the static phase. Within each 24 h segment, the environmental temperature trend was nearly identical, which indicated that the experimental results were not affected by significant environmental temperature fluctuations. Figure 8 shows the trend in the evaporative mass flow rate over the static phase. During the initial precooling phase, the mass flow rate rapidly increased until the flow meter was connected to the system after 60 h of pre-cooling. As shown in Figure 8, the mass flow rate reached its peak at 69 h and subsequently decreased rapidly before stabilizing after 72 h. Throughout this stable phase, minor fluctuations in the mass flow rate coincided with minor variations in the environmental temperature, but overall, the temperature followed an initial increasing and then decreasing pattern, peaking at 101 h.

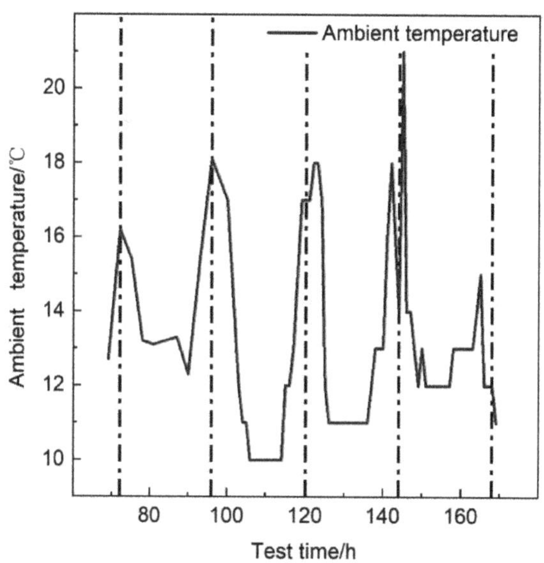

Figure 7. Trends in environmental temperature variations.

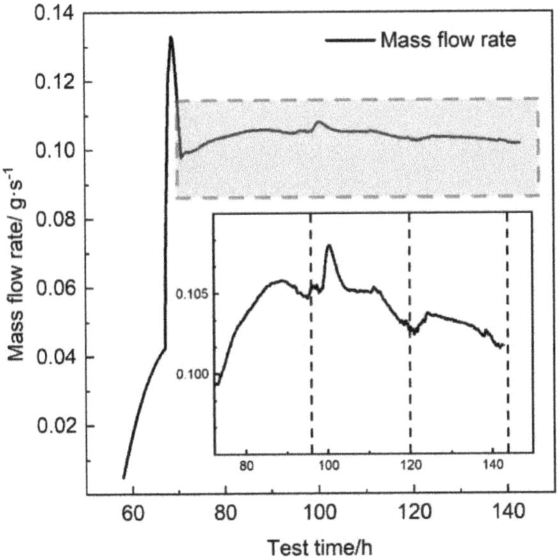

Figure 8. Variation in the amount of helium gas that evaporated from the liquid helium Dewar.

Figure 8 shows the variation in the amount of helium gas that evaporated from the liquid helium Dewar. Throughout the process of liquid helium evaporation, the amount of helium that evaporated was influenced by the fill level and by the environmental temperature. Previous research indicates varying degrees of relative increased container heat leakage between the gas and liquid phases [21]. As the fill level decreased, there was an increase in the contact area between the gas phase and the container walls, which enhanced natural convection and increased container heat leakage. This effect peaked at 101 h (as shown in Figure 8), with a fill level of 57.19%, corresponding to the maximum evaporation rate. As the volume of the gas phase continued to expand, the amount of heat transfer through conduction decreased because of the higher overall temperature of the gas phase than the liquid phase, which resulted in decreased heat leakage and a reduction in the evaporation rate.

On the basis of the data in Figure 8, the daily boil-off rates were calculated via the following equation:

$$\alpha_0 = \frac{q_{mg}}{\rho_0 V} \times 100\% \tag{7}$$

The daily boil-off rates were 14.3% from 72 to 96 h, 14.5% from 96 to 120 h, and 14.2% from 120 to 144 h. The measurement error in the daily boil-off rate over three days was within 5%, validating the accuracy of the data measurements.

4.3. Heat Leakage from the Experimental Liquid Helium Dewar

Figure 9 shows the temperature trend for the sensor T01 at the neck tube, whereas Figure 10 depicts the temperature distribution for sensors T02 and T03 along the neck tube over the duration of the static phase. As illustrated in Figure 9, T01 showed significant temperature fluctuations at 50 h due to helium Dewar reliquification, and the temperature stabilized after 72 h and exhibited a continuous upward trend within a narrow range. In Figure 10, it can be seen that the temperatures at sensors T02 and T03 at the neck tube rapidly decreased in the initial precooling static phase and showed substantial fluctuations during the nonequilibrium phase at 72 h. Notably, the sharp temperature increases and decreases corresponded to the liquid helium reliquification intervals, and the temperature stabilized gradually after 72 h, with temperatures at the neck tube rising continuously as the liquid level inside the Dewar decreased.

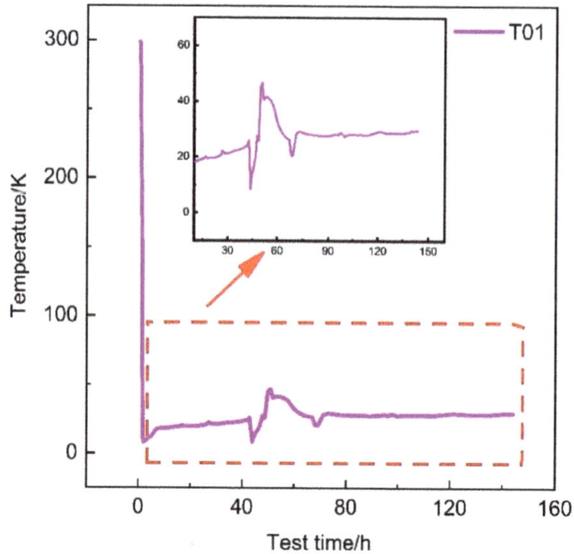

Figure 9. Variation in T01 at the neck tube.

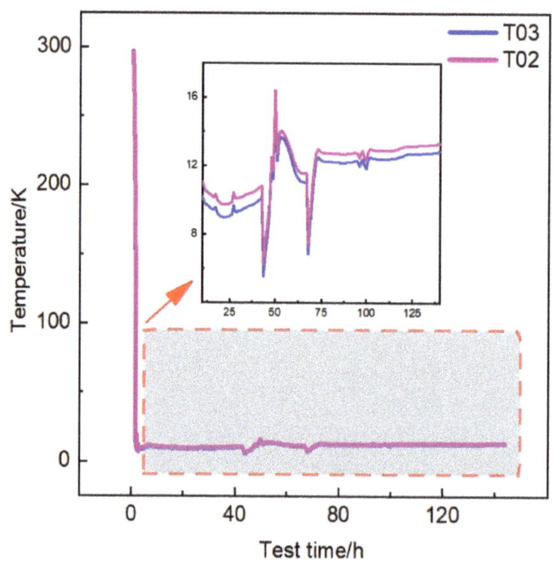

Figure 10. Variation in T02–T03 at the neck tube.

Thermal Analysis of Dewar

Based on the data from Figures 8–10, both temperature and flow rate exhibit slight variations during the 72–144 h testing period, but generally stabilize, indicating that a steady-flow condition is reached during this interval. Temperature data shows a distinct temperature gradient in the gas phase, with the gradient inside the neck tube being significantly steeper than that within the container. Therefore, the following assumptions are made for the thermal analysis in this section:

1. The pressure remains constant throughout the test, and steady-flow conditions are achieved during the 72–144 h period.
2. Heat leakage is assumed to solely originate from the surrounding environment.
3. The liquid phase is treated as an incompressible fluid, maintaining a saturated and uniform phase.
4. While the temperature gradients in the gas phase within the container and neck tube differ, both are assumed to follow a linear distribution.

As shown in Figure 11, Q_l is the heat leakage in the liquid phase; Q_g is the heat leakage in the vapor phase; Q_{nt} is the heat leakage in the neck tube. During the testing process, heat is transferred from the surrounding environment to the liquid and gas inside cryogenic vessel, resulting in the evaporation of a portion of the liquid into gas. This leads to the formation of vapor at the liquid–gas interface. The vapor then exits the container, and its volumetric flow rate is measured using a mass flow meter. The mass and energy equations for both the gas and liquid phases are expressed as follows:

$$\frac{d(\rho_g V_g)}{dt} + \frac{d(\rho_l V_l)}{dt} = -\frac{d(\rho_v V_v)}{dt} \tag{8}$$

$$\frac{d(\rho_g V_g e_g)}{dt} + \frac{d(\rho_l V_l e_l)}{dt} = Q_W - \frac{d(\rho_v V_v)}{dt} h_v \tag{9}$$

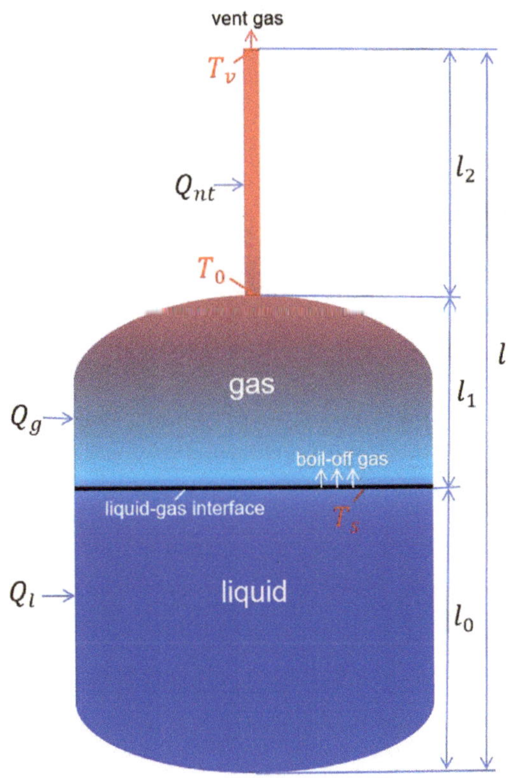

Figure 11. Diagram of physical model.

Under the assumptions of incompressible liquid and steady flow conditions, Equations (8) and (9) can be rewritten as follows:

$$\frac{d(\rho_g V_g)}{dt} + \rho_l \frac{dV_l}{dt} = -\rho_v \frac{dV_v}{dt} \tag{10}$$

$$\rho_g V_g \frac{de_g}{dt} + e_g \frac{d(\rho_g V_g)}{dt} + \rho_l V_l \frac{de_l}{dt} + \rho_l e_l \frac{dV_l}{dt} = Q_W - \rho_v \frac{dV_v}{dt} h_v \tag{11}$$

As the evaporated liquid is produced at the interface, its volume is occupied by the gas near the interface, where the gas temperature is equal to the saturated liquid temperature. Therefore, their relationships can be expressed as follows:

$$-\rho_l \frac{dV_l}{dt} = \rho_v \frac{dV_v}{dt} + \frac{\rho_{sg}}{\rho_l} \cdot (-\rho_l \frac{dV_l}{dt}) \tag{12}$$

It is also known that internal energy is a single-valued function of temperature. It can be expressed as follows:

$$e = cT \tag{13}$$

By combining Equations (10)–(13), the energy equation can be rewritten as follows:

$$\rho_g V_g c_g \frac{dT_g}{dt} + \rho_l V_l c_l \frac{dT_l}{dt} = Q_W - \rho_v \frac{dV_v}{dt} (h_v - e_g + \frac{\rho_l}{\rho_l - \rho_{sg}} (e_g - e_l)) \tag{14}$$

Based on the assumption of steady-state flow, the temperature is considered constant over time. Therefore, the expression for heat leakage is as follows:

$$Q_W = \rho_v \dot{V}_v (h_v - e_g + \frac{\rho_l}{\rho_l - \rho_{sg}} (e_g - e_l)) \tag{15}$$

e_g is the average specific internal energy in the gas phase. As illustrated in Figure 11, assuming a linear temperature distribution, the temperature distribution in the gas phase is as follows:

$$T_g = \begin{cases} \frac{T_0 - T_s}{l_1}(l - l_0) + T_s & l_0 < l \leq l_0 + l_1 \\ \frac{T_v - T_0}{l_2}(l - l_0 - l_1) + T_0 & l_0 + l_1 < l \leq l_0 + l_1 + l_2 \end{cases} \tag{16}$$

The cross-sectional area of the container is denoted as A_1, and the cross-sectional area of the neck tube is denoted as A_2. By combining Equations (13), (15) and (16), the relationship between the heat leakage, temperature, and flow rate is expressed as follows:

$$Q_W = \rho_v \dot{V}_v (h_v - c_g (\frac{T_0}{2} + \frac{T_s A_1 l_1 + T_v A_2 l_2}{2(A_1 l_1 + A_2 l_2)}) + \frac{\rho_l}{\rho_l - \rho_{sg}} \left(c_g (\frac{T_0}{2} + \frac{T_s A_1 l_1 + T_v A_2 l_2}{2(A_1 l_1 + A_2 l_2)}) - c_l T_l \right)) \tag{17}$$

In practical applications, $A_1 l_1 \gg A_2 l_2$, thus the above expression simplifies to:

$$Q_W = \rho_v \dot{V}_v (h_v - c_g \frac{T_0 + T_s}{2} + \frac{\rho_l}{\rho_l - \rho_{sg}} \left(c_g \frac{T_0 + T_s}{2} - c_l T_l \right)) \tag{18}$$

If $T_a = \frac{T_0 + T_s}{2}$, the expression can be rewritten as follows:

$$Q_W = \rho_v \dot{V}_v (h_v - e_{g,a} + \frac{\rho_l}{\rho_l - \rho_{sg}} (e_{g,a} - e_l)) \tag{19}$$

As shown in Figure 11, the heat leakage in the Dewar can be divided into two components: the heat leakage from the inner container and the neck tube. Based on Equation (19), where $T_v = T01$ and $T_0 = T03$, the expression for the total heat leakage is given as follows:

$$Q_{ic} = \rho_{01}\dot{V}_v(h_{03} - e_{g,a} + \frac{\rho_l}{\rho_l - \rho_{sg}}(e_{g,a} - e_l)) \tag{20}$$

$$Q_{nt} = \rho_{01}\dot{V}_v(h_{01} - h_{03}) \tag{21}$$

On the basis of the NIST database and experimental data, Q_{ic} and Q_{nt} were calculated at different times, starting from the 72 h static phase, as shown in Figure 12. It can be observed that Q_{ic} and Q_{nt} fluctuate within a small range; however, it is evident that the heat leakage from the neck tube is significantly more than that from the container.

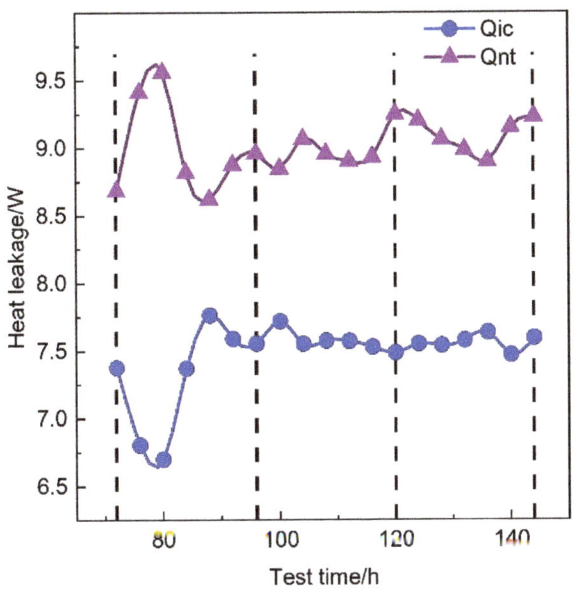

Figure 12. Variation in heat leakage from the liquid helium Dewar.

Based on the above, it can be observed that the heat leakage through the neck tube is significant, which in turn affects the temperature distribution within the neck tube, specifically the value of T_0 shown in Figure 11. This indirectly influences the evaporation rate of the inner container. Assuming that the heat leakage of the inner container is constant, the relationship between the heat leakage through the neck tube and the evaporation rate, as described by Equations (20) and (21), is shown in Figure 13.

As shown in Figure 13, the heat leakage through the neck tube influences the evaporation rate in a linear fashion. Therefore, the design of the neck tube is of critical importance. The heat leakage from the inner container is approximately 7.5 W, with a heat flux of 2.254 W/m^2.

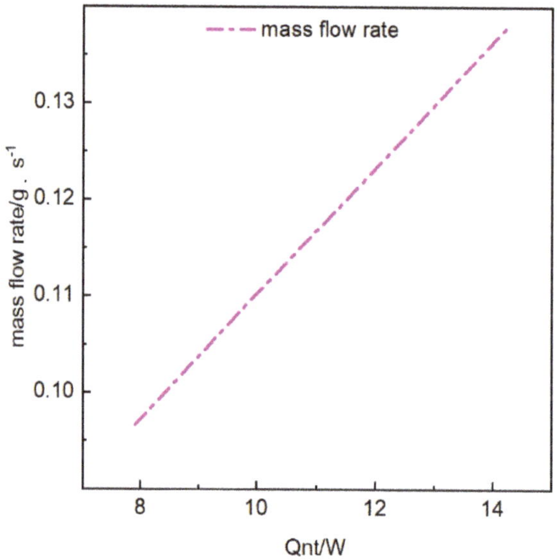

Figure 13. Relationship between evaporation and heat leakage of neck tube.

4.4. Thermal Insulation Characteristics of MLI

Figure 14 displays the temperature variations between layers of the MLI as recorded by the sensors T11–T16. During the initial 0–72 h of the precooling static phase, T11 rapidly decreased within a few hours and stabilized thereafter, whereas T16 stabilized only after 50 h of the static phase. Once stabilized, T15 and T16 exhibited minimal temperature fluctuations, whereas T11, T12, T13, and T14 exhibited noticeable variations, which were correlated with environmental temperature changes, eventually stabilizing. Based on the average temperatures every 24 h during the stable phase, the trends in the MLI temperatures at 72–96 h, 96–120 h, and 120–144 h of the experiments are presented in Figure 13. Notably, the temperature at the outer wall of layer 0 of the insulation was equivalent to 7.4 K; however, this temperature was calculated on the basis of an actual experimental pressure of 6.7 psia, which should correspond to a saturated liquid helium temperature of 4.63 K. Therefore, there was a contact thermal resistance between the sensor and the inner container, referred to as the self-heating effect. The errors caused by the self-heating effect diminish with increasing temperature; thus, in subsequent calculations, a temperature correction of 4.63 K is only required for the layer 0.

Using a generalized mathematical model for MLI, specifically the layer-and-layer model [22,23], numerical simulations of temperature among MLIs were conducted and compared with experimental results.

Layer-and-layer model:

$$q_{total} = q_s + q_g + q_r \tag{22}$$

Radiative heat transfer:

$$q_r = \frac{\sigma(T_h^4 - T_c^4)}{(\frac{1}{\varepsilon_h} + \frac{1}{\varepsilon_c} - 1)} \tag{23}$$

Residual gas conduction:

The formulas for calculating Kn are presented in Equations (24) and (25).

$$Kn = \frac{l}{L} \tag{24}$$

$$p = 6.44 \times 10^3 \frac{\eta}{l} \sqrt{\frac{T'}{M}} \tag{25}$$

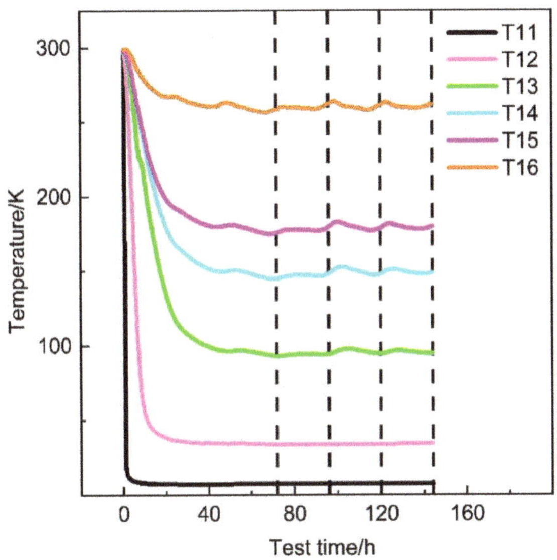

Figure 14. Temperature variations between multilayer insulation according to sensors T11–T16.

Thus, $Kn > 10$, which can be calculated according to experimental conditions. The residual gas is in the free molecular regime. The gas conduction is calculated by Equation (26).

$$q_g = \frac{\gamma + 1}{\gamma - 1} \sqrt{\frac{R}{8\pi M T'}} p\alpha(T_h - T_c) \tag{26}$$

Solid conduction:

$$q_s = \frac{X f \lambda}{D}(T_h - T_c) \tag{27}$$

For Dacron net, the relationship between λ_{dn} and T is given by:

$$\lambda_{dn} = 0.017 + 7 \times 10^{-6} \times (800 - T') + 0.0228\ln(T') \tag{28}$$

For hollow glass microspheres, the relationship between λ_{HGMs} and T' is described by Equation (29) [24]:

$$\lambda_{HGMs} = 0.9885 \times 10^{-6} T' \tag{29}$$

Figure 15 shows a comparison between the theoretical simulation results and experimental results of the temperature distribution across the MLI system. Under identical boundary conditions (warm temperature of 260 K and cold boundary temperature of 4.63 K), the overall relative error between the MLI and HGM temperature distributions in the simulation results and experimental findings is 14.3%, with a maximum deviation of 22.3%. The simulation results showed a trend of decreasing temperature gradients between layers 24 to 60 of the MLI, whereas the experimental results exhibited fluctuations in the temperature gradients. In the previous literature, researchers extensively discussed theoretical studies on MLI, often suggesting that temperature gradients within insulation layers decrease as temperatures increase. However, the experimental results indicate that temperature gradients within the insulation layers fluctuate with increasing temperatures.

To explain this phenomenon, the apparent thermal conductivity distribution of the insulation layers was calculated on the basis of experimental data at the midpoint of the 1/2 MLI layer, which was located at a cylindrical radius with a curvature of 0.495 m. According to Fourier's law, the apparent thermal conductivity can be calculated via Equation (30):

$$\lambda_a = \frac{\Phi \ln\left(\frac{r_h}{r_c}\right)}{2\pi \bar{l}(T_h - T_c)} \tag{30}$$

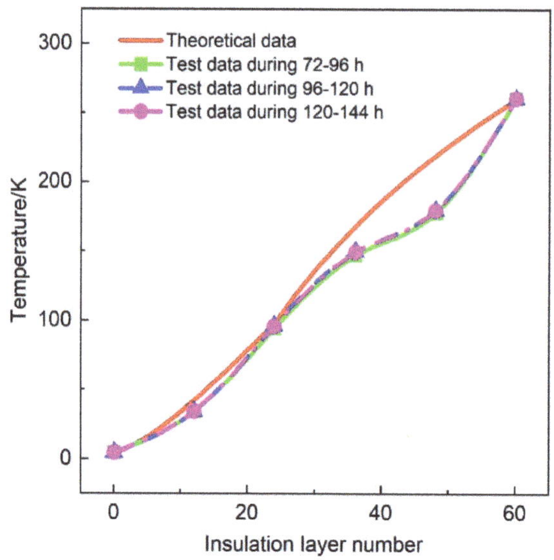

Figure 15. Comparison between theoretical simulations and experimental results of the temperature distribution across the MLI system.

On the basis of the equation above, the apparent thermal conductivity of the 60-layer MLI with glass microspheres was 2.887×10^{-4} W/(m·K). To better reflect the variation in apparent thermal conductivity with the number of layers, segmented calculations of apparent thermal conductivity were conducted on the basis of the experimental data.

As shown in Figure 16, layers 0 to 60 were divided into five regions labeled I to V. The regions I to V are depicted in Figure 2. The apparent thermal conductivity for each region was calculated on the basis of the interlayer temperature. The temporal variation in apparent thermal conductivity between layers of MLI is depicted in Figure 16a. Region I had the highest values, whereas Region V had the lowest values. In contrast, Regions II, III, and IV displayed notable fluctuations in apparent thermal conductivity, which correlated with variations in ambient temperature.

To further illustrate the trend in apparent thermal conductivity with increasing layer number, the average temperatures and flow rates from 72 to 96 h, 96 to 20 h, and 120 to 144 h were taken as sample points. The computed values for Region I represent the apparent thermal conductivity between the 0–6th layers, and this process continues sequentially. Figure 16b shows the variation in apparent thermal conductivity with increasing interlayer temperature, which significantly fluctuated as the layer number (and temperature) increased. The apparent thermal conductivity reached a value of 1.721×10^{-4} W/(m·K) in the temperature range of 4.62–34.2 K and 5.342×10^{-4} W/(m·K) in the range of 179–260 K.

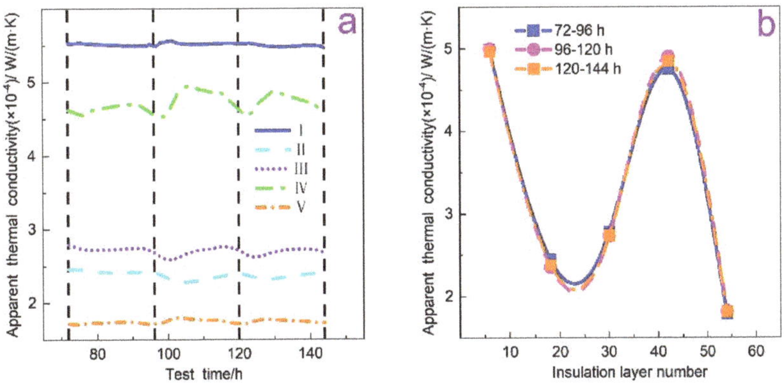

Figure 16. Apparent thermal conductivity of multilayer insulation as a function of (**a**) test time and (**b**) the number of insulation layers.

In the theoretical analysis of multilayer insulation, the heat transfer coefficient varies with temperature, indicating that the apparent thermal conductivity of MLIs is a function of temperature. The experimental results indicated that the use of MLI containing hollow glass microspheres as spacers resulted in fluctuations in the apparent thermal conductivity with increasing temperature. According to the expressions for solid heat transfer and radiative heat transfer, both monotonically increase with temperature. Conversely, under high vacuum, on the basis of rarefied gas theory [25], the gas thermal conductivity decreases with increasing temperature. Therefore, in the MLI heat transfer process, the gas conduction term dominates in the thermal insulation layers 0 to 24, resulting in an increasing temperature gradient. In layers 24 to 36, solid conduction becomes predominant, leading to a decreasing temperature gradient. In this range, the trends observed in both simulation and experiment are consistent. In the outer thermal insulation layers, the experimental data reveal a contrasting trend to simulation, which could be attributed to the dominance of the gas conduction term instead. The observed changes in the dominant term are likely attributable to an uneven distribution of pressure. Although direct measurement of pressure within the insulation layer is challenging, experimental temperature data and extensive theoretical research on MLI heat transfer mechanisms provide substantial evidence to support this assertion.

5. Conclusions

In the experiment, MLI with hollow glass microspheres (HGMs) as spacers and liquid helium as the working fluid were used to study the thermal insulation performance of a liquid helium Dewar system. The following conclusions were drawn:

(1) The experimental system utilizing the MLI–HGMs presented an average daily boil-off rate of 14.4%, heat leakage of 6.6 W, and a heat flux of 2.254 W/m^2 from the inner container.

(2) A theoretical thermal analysis of the experimental Dewar was performed, leading to the derivation of a relationship between the heat leakage, evaporation rate, enthalpy, and internal energy. The heat leakage of the neck tube has a significant effect on the evaporation.

(3) The uncertainty of the temperature sensors is 1.16%, while the comprehensive measurement uncertainty of the system is 2.78%. There exists a certain degree of discrepancy between the simulated and experimental results of the interlayer temperature

distribution of the insulation layer, with an overall error of 14.3% and a maximum error of 22.3%.

(4) The apparent thermal conductivity of different layers of MLI is temperature-dependent, as supported by extensive literature indicating a unidirectional increase in apparent thermal conductivity with increasing temperature. In this system, experimental analysis revealed that the overall apparent thermal conductivity of the 60-layer MLI, operating at a cold end of 4.63 K and an ambient temperature of 286 K, was 2.887×10^{-4} W/(m·K). The apparent thermal conductivity between the layers of insulation exhibited significant fluctuations with temperature, ranging from a maximum of 5.342×10^{-4} W/(m·K) to a minimum of 1.721×10^{-4} W/(m·K).

(5) During the heat transfer process in MLI with HGMs, the dominant heat transfer mechanisms are likely to vary with temperature. This phenomenon is potentially associated with the non-uniform distribution of pressure following complete cooling.

In this work, passive insulation performance studies were successfully conducted for the test system in the liquid helium temperature range. In future studies, the system will incorporate an external GM cooler to re-liquefy the evaporated helium gas and return it to the Dewar. The focus will be on investigating the impact of the active cooling structure on the flow field, particularly the condensation efficiency, energy consumption, and pressure variations in the internal fluid at different cold head positions. This research is of significant importance for the zero-evaporation storage of liquid helium.

Author Contributions: Methodology, Y.C., L.G. and X.X.; Validation, Y.C.; Investigation, Y.C., L.G., Q.J., W.Z. and P.W.; Resources, X.X.; Writing—original draft, Y.C.; Writing—review & editing, Y.C., L.G. and Q.J.; Visualization, Y.C.; Funding acquisition, X.X. All authors have read and agreed to the published version of the manuscript.

Funding: This work was supported by the fund of the Strategic Pilot Science and Technology Project of the Chinese Academy of Sciences (XDC10010300).

Data Availability Statement: The original contributions presented in this study are included in the article. Further inquiries can be directed to the corresponding author.

Conflicts of Interest: The authors declare no conflict of interest.

Nomenclature

Nomenclature

Q	heat leakage, W
$\overline{A}$	equivalent area,m^2
L	length of neck tube, m
λ	thermal conductivity, W/(m·K)
$\bar{l}$	equivalent length, m
r	curvature radius of multilayer insulation, m
T	absolute temperature, K
q_m	mass flow rate, kg/s
q_{mg}	daily average mass flow rate, kg/d
V	inner container volume
α_0	daily boil-off rate
q_0	volumetric flow rate under standard conditions, m^3/s
ρ_0	density of helium gas under standard conditions, kg/m^3
h_v	enthalpy of helium vent gas under actual conditions, J/kg
c	specific heat capacity at constant pressure of helium, J/(kg·K)
e	enthalpy of cold helium gas leaving neck tube under actual conditions, J/kg

Q_{ic}	heat leakage of inner container, W
Q_{nt}	heat leakage of neck tube, W
ε	effective emissivity
η	dynamic viscosity
l	molecular free path
Kn	Knudsen number
γ	specific heat ratio related to the gas between layers, 1.4 for air
R	gas constant, 8.314 J/(mol·K)
σ	Stefan-Boltzmann constant, 5.67×10^{-8} W/$\left(\mathrm{m}^2\cdot\mathrm{K}^4\right)$
M	molar mass related to gas between layers, for air, 29 g/mol
T'	characteristic temperature between layers, K
p	characteristic pressure between layers, pa
α	constant related to gas
X	constant related to multilayer insulation
f	spacer sparsity of multilayer insulation
D	distance between layers
λ_a	apparent thermal conductivity, W/(m·K)
Φ	total heat flux, W
PFD	process flow diagram
Subscripts	
nt	neck tube
il	insulation layer
g	gas
h	hot boundary
c	cold boundary
rc	radiation conduction
gc	gas conduction
sc	solid conduction
dn	Dacron net

References

1. Van Sciver, S.W.; Timmerhaus, K.D.; Clark, A.F. *Helium Cryogenics*; Springer: New York, NY, USA, 2012.
2. Balibar, S. Laszlo Tisza and the two fluid model of superfluidity. *Comptes Rendus. Phys.* **2017**, *18*, 586–591.
3. Hastings, L.J.; Plachta, D.W.; Salerno, L.; Kittel, P. An overview of NASA efforts on zero boiloff storage of cryogenic propellants. *Cryogenics* **2001**, *41*, 833–839.
4. Tseng, C.; Yamaguchi, M.; Ohmori, T. Thermal conductivity of polyurethane foams from room temperature to 20 K. *Cryogenics* **1997**, *37*, 305–312.
5. Liu, Z.; Li, Y.; Xie, F.; Zhou, K. Thermal performance of foam/MLI for cryogenic liquid hydrogen tank during the ascent and on orbit period. *Appl. Therm. Eng.* **2016**, *98*, 430–439. [CrossRef]
6. Allen, M.S.; Baumgartner, R.G.; Fesmire, J.E.; Augustynowicz, S.D. Advances in microsphere insulation systems. *AIP Conf. Proc.* **2004**, *710*, 619–626.
7. Naes, L.G., Jr.; Dammann, R. Multilayer insulation performance at low cold boundary temperatures. In Proceedings of the Cryogenic Optical Systems and Instruments IX, Seattle, WA, USA, 8 July 2002; Volume 4822, pp. 94–103.
8. Vanderlaan, M.; Stubbs, D.; Ledeboer, K.; Ross, J.; Van Sciver, S.; Guo, W. Repeatability Measurements of Apparent Thermal Conductivity of Multilayer Insulation (MLI). *IOP Conf. Ser. Mater. Sci. Eng.* **2017**, *278*, 012195. [CrossRef]
9. Shu, Q.S.; Fast, R.W.; Hart, H.L. Heat flux from 277 to 77 K through a few layers of multilayer insulation. *Cryogenics* **1986**, *26*, 671–677. [CrossRef]
10. Sun, P.J.; Wu, J.Y.; Zhang, P.; Xu, L.; Jiang, M.L. Experimental study of the influences of degraded vacuum on multilayer insulation blankets. *Cryogenics* **2009**, *49*, 719–726. [CrossRef]
11. Bapat, S.L.; Narayankhedkar, K.G.; Lukose, T.P. Experimental investigations of multilayer insulation. *Cryogenics* **1990**, *30*, 711–719. [CrossRef]
12. Hastings, L.J.; Martin, J.J. Experimental testing of a foam/multilayer insulation (FMLI) thermal control system (TCS) for use on a cryogenic upper stage. *AIP Conf. Proc.* **1998**, *420*, 331–341.

13. Hastings, L.J.; Hedayat, A.; Brown, T.M. Analytical Modeling and Test Correlation of Variable Density Multilayer Insulation for Cryogenic Storage. 2004. Available online: https://ntrs.nasa.gov/api/citations/20040121015/downloads/20040121015.pdf (accessed on 16 December 2024).
14. Zheng, J.; Chen, L.; Wang, J.; Xi, X.; Zhu, H.; Zhou, Y.; Wang, J. Thermodynamic analysis and comparison of four insulation schemes for liquid hydrogen storage tank. *Energy Convers. Manag.* **2019**, *186*, 526–534.
15. Zhang, C.; Li, C.; Jia, W.; Pang, Y. Thermodynamic study on thermal insulation schemes for liquid helium storage tank. *Appl. Therm. Eng.* **2021**, *195*, 117185.
16. Zheng, J.; Chen, L.; Wang, P.; Zhang, J.; Wang, J.; Zhou, Y. A novel cryogenic insulation system of hollow glass microspheres and self-evaporation vapor-cooled shield for liquid hydrogen storage. *Front. Energy* **2020**, *14*, 570–577. [CrossRef]
17. Wawryk, R.; Rafałowicz, J. Heat transfer in microsphere insulation. *J. Therm. Anal. Calorim.* **1988**, *34*, 249–257.
18. Wang, P.; Liao, B.; An, Z.; Yan, K.; Zhang, J. Measurement and calculation of cryogenic thermal conductivity of HGMs. *Int. J. Heat Mass Transf.* **2019**, *129*, 591–598.
19. Fesmire, J.E.; Augustynowicz, S.D. Thermal performance testing of glass microspheres under cryogenic vacuum conditions. *AIP Conf. Proc.* **2004**, *710*, 612–618.
20. Rugaiganisa, B.M.; Yoshihara, T.; Yoshiwa, M.; Nakagawa, S.; Hirai, A. Experimental investigation on heat leak into a liquid helium Dewar. *Cryogenics* **1990**, *30*, 942–946.
21. Zuo, Z.; Jiang, W.; Qin, X.; Huang, Y. A numerical model for liquid–vapor transition in self-pressurized cryogenic containers. *Appl. Therm. Eng.* **2021**, *193*, 117005. [CrossRef]
22. McIntosh, G.E. Layer by layer MLI calculation using a separated mode equation. In *Advances in Cryogenic Engineering*; Springer: Boston, MA, USA, 1994; pp. 1683–1690.
23. Yin, L.; Yang, H.; Ju, Y. Review on the key technologies and future development of insulation structure for liquid hydrogen storage tanks. *Int. J. Hydrogen Energy* **2024**, *57*, 1302–1315. [CrossRef]
24. Wang, P.; Ji, L.; Yuan, J.; An, Z.; Yan, K.; Zhang, J. Modeling and optimization of composite thermal insulation system with HGMs and VDMLI for liquid hydrogen on orbit storage. *Int. J. Hydrogen Energy* **2020**, *45*, 7088–7097. [CrossRef]
25. Sundén, B.; Fu, J. *Heat Transfer in Aerospace Applications*; Academic Press: Cambridge, MA, USA, 2016.

Disclaimer/Publisher's Note: The statements, opinions and data contained in all publications are solely those of the individual author(s) and contributor(s) and not of MDPI and/or the editor(s). MDPI and/or the editor(s) disclaim responsibility for any injury to people or property resulting from any ideas, methods, instructions or products referred to in the content.